Underwater California

by WHEELER J. NORTH

With sections on underwater photography

by Robert Hollis

UNIVERSITY OF CALIFORNIA PRESS
BERKELEY LOS ANGELES LONDON

California Natural History Guides
Arthur C. Smith, General Editor

University of California Press
Berkeley and Los Angeles, California

University of California Press, Ltd.
London, England

ISBN: 0-520-03025-7 (clothbound)
0-520-03039-7 (paperbound)
Library of Congress Catalog Card Number: 75-13153
Printed in the United States of America

CONTENTS

Chapter 1

INTRODUCTION

This book is concerned primarily with the living creatures of underwater California. Our principal objective is to enhance enjoyment of the submarine world by providing a better understanding about its inhabitants. The knowledgeable diver has a better time than the individual who knows very little about what he sees. Seaweeds, for example, are not mere entangling nuisances to the knowledgeable diver. They are viewed as homes for countless animals, food for many others, and indispensable shelters for the young of a great many swimming forms such as fishes. Sea life observed or collected for the sake of curiosity has attraction; but when you know what the organisms are, what they do for a living, and the relationships they bear to their environment, the depth and duration of your interest approaches total fascination.

The latter part of this book lists the more common plants and animals of underwater California. For some obscure reason, the human mind finds objects much more interesting if they have names (for example, a baseball game is more fun if you can name the players). Your favorite diving spots will become even more enchanting if you take the trouble to identify and learn the names of the common organisms present.

The way of life in the sea is profoundly influenced by the character of the environment. We have therefore included pertinent aspects of marine ecology and oceanography, particularly those factors that influence distribution of plants and animals. Residents of the ocean are often choosy about their homes, and if you are familiar with the environmental requirements of a species it is usually much easier to find it. Wherever possible we have also included information on natural history as a guide to finding and collecting. Considerable attention has been given to the geography of underwater California. In many instances the distribution of plants and animals is more easily understood in the light of geographical knowledge; likewise, some locations have more to offer than others. We have chosen a few of the more outstanding submarine paradises for detailed description. Many equally beautiful areas were omitted for lack of space; undoubtedly many more remain completely unknown and await discovery.

We have also included a short section on pertinent helpful techniques which may not have been covered in a basic training course for divers. Otherwise we have assumed that the reader has already mastered the fundamentals of diving. Elementary aspects of biology were also omitted. Additional information from other sources would be helpful to the reader who is not familiar with the phyla of the plant and animal kingdoms or with ecological concepts such as communities and food webs. Companions to this volume, *Introduction to Seashore Life of the San Francisco Bay Region and the Coast of Northern California* by Joel W. Hedgpeth, *Seashore Plants of Northern California* and *Seashore Plants of Southern California* by E. Yale Dawson, *Marine Mammals of California* by Robert T. Orr, *Marine Food and Game Fishes of California* by John E. Fitch and Robert J. Lavenberg, and *Seashore Life of Southern California* by Sam Hinton, contain most of the pertinent fundamentals. A person without diving experience is strongly urged to seek instruction from a competent school for beginners. Almost any store that sells underwater equipment can direct you to a source of good instruction.

ACKNOWLEDGMENTS

It is a pleasure to acknowledge invaluable assistance from Wendy Bergman, Marjorie Connely, Karen Ehlbeck, Joyce Hsiao, Virginia Kuga, and Ann Long in preparing and typing the manuscript. Photo credits to Rich Bergero for Plates 2c, 5f, 6b and d, and 8a; Robert Hollis for Plates 1a and d, 2b, 3a and c, 4b and f, 7a, b, and c, and 8b and e; Charles Nicklin, Jr. for Plates 1b and c, 2d, e, and f, 3b, 4a, c, d, and e, 5a, 8c, d, and f, and the cover photo. All other photos are by the author.

Chapter 2

THE CALIFORNIA COAST

HISTORICAL ASPECTS

The California coast displays a virtually infinite variety of underwater environments. Almost every beach, cove, and inlet is strange and different, fascinatingly unique. This chapter discusses the major climatic influences that act on these multitudinous environments, conferring enchanting characteristics on each.

The recent past is an important key to understanding the present. Ocean conditions, particularly water temperatures, have not been constant over the past few thousand years. Carl L. Hubbs of the Scripps Institution of Oceanography has shown that major fluctuations in average temperature have occurred, in cycles lasting hundreds of years. Cold-loving animals that do not presently occur south of central California formerly inhabited the coast as far south as Cape San Lucas at the tip of Baja California. About 400 years ago the ocean climate passed through a minimum temperature phase and has been slowly warming ever since. During warming phases the cold-loving species presumably recede northward and are replaced by tropical or semitropical species extending their ranges from the south. The reverse occurs during cooling portions of a cycle.

There is seldom any sharp boundary separating cold- from warm-loving species. The uneven geography of the coast allows "cold spots" and "warm spots" to persist at considerable distances from where such temperatures are normal for the area. Consequently we find "islands" of cold- or warm-loving organisms inhabiting such spots. Presumably these plants and animals are descendants of creatures formerly distributed more generally, when the temperature of the entire region corresponded to their requirements.

Even more astounding, sea level has fluctuated substantially in recent times. In southern California 30,000 years ago, waves broke on beaches that were about 400 feet (122m) below the present level. Subsequently, mean sea level has risen and fallen. About 7000 years ago the edge of the sea stood somewhat more than 100 feet (30.5m) below our modern shoreline. Since then sea level is believed to have risen fairly uniformly. Artifacts left by California Indians are now often found at depths of 30 to 40 feet (9-12m). Common geological features of our topside shores such as ancient coastal plains, former riverbeds, stream-eroded canyons, and wave-cut terraces terminating in cliffs are also plentiful in shallow water under the sea's surface.

PRINCIPAL FEATURES

The offshore bottom, out to depths of about 600 feet (183m), is called the *continental shelf* (Fig. 1). Of chief interest to divers is the portion of the shelf lying between the intertidal region and approximately the 100-foot (30.5m) depth contour. A few areas display a sufficient diversity and abundance of life below this depth to merit attention, but safety considerations limit effective use of scuba equipment to maximum depths of about 200 feet (61m). We will call this region the *scuba zone*.

Life abounds both in the water and on the bottom. Organisms associated primarily with the water well above the bottom are called *pelagic*. Organisms preferring space on or near the bottom are called *benthic*. The pelagic division is divided by an imaginary line at the edge of the continental shelf into the *neritic province* (nearest shore) and the *oceanic province*. The benthic division has been subdivided several ways by different authors. For our purposes we will define three main regions: the *intertidal* lying between high and low tides; the *shallow littoral* from the low-tide level to the region where attached plants become sparse (generally at depths of

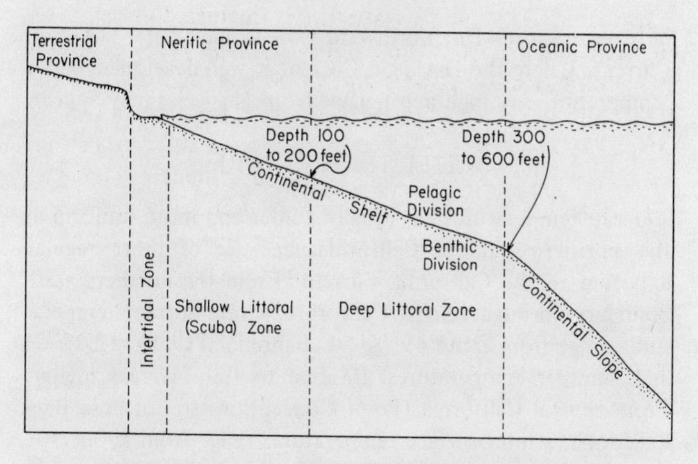

Fig. 1. Cross-sectional view of the ocean and sea floor near the shore, showing provinces, divisions, zones, and outermost depths.

50 to 150 feet [15-45.5m]); and the *deep littoral* extending seaward to the edge of the shelf.

When investigating the pelagic division along California, the diver usually encounters neritic organisms near the main coast and oceanic organisms in the vicinity of the islands. The benthos of the mainland and the islands is generally similar, although there are species unique to each situation.

WATER CHARACTERISTICS

Most of the water that bathes California shores has been brought from the Gulf of Alaska by the southward-flowing California Current. During its journey the portions nearest the continent receive fresh water from the rivers of Alaska, Canada, and the northwestern United States. These additions do not affect the overall *salinity* (salt concentration) significantly, but substantial amounts of trace minerals enter the sea and contribute to plant nutrition. Drainage from the land decreases markedly south of San Francisco. At times the southward flow of the California Current is reversed, particularly near shore. This northward flow is named the Davidson Current. Where the Davidson Current is well developed, water temperatures are high and pelagic tropical species may appear.

WATER TEMPERATURE

Ocean temperatures are usually colder and more uniform in the northern part of California, because of more regular exposure to the California Current. From the northern state boundary through central California, winter surface temperatures range from about 49° to 54° Fahrenheit (9.44°-12.22°C) and summer temperatures are one to four degrees higher. From central California (Point Conception) to northern Baja California, winter surface temperatures range from about 50° to 55° Fahrenheit (10°-12.78°C) becoming 70° to 74° (21.11°-22.33°C) for usual summer conditions. Temperature appears to be extremely important, strongly influencing species distribution. Forms that we will classify as cold-water species usually inhabit regions where temperatures do not exceed 55° to 60° (12.78°-15.56°C) for appreciable periods of time. Endemic species characteristic of warm water can usually tolerate substantial periods of cold. Consequently, warm-water species at times survive in the coldest California sea water.

TURBIDITY

A simple measure of turbidity is the distance that can be seen underwater (sometimes confused with *turbulence*, which is random swirling motion in the water). Water absorbs light and this affects underwater visibility, but *turbidity* refers to the interference with light by tiny particles suspended in the water. These particles may be nonliving mineral specks or they may be tiny microorganisms (algae, fungi, protozoa, and bacteria).

Nonliving particles such as sand and silt are more abundant near shore, especially in the vicinity of rivers and bays. Strong water movements from currents, waves, and wind tend to bring inanimate particles into suspension from the bottom.

Suspended microbes usually are also more abundant near shore. Their growth is encouraged by nutrients from rivers, bays, and discharged wastes. Because both of greater runoff from land and of rougher seas in northern California, turbidity is usually greater along this part of the coast. Turbidity usually decreases with increasing distance from shore and with greater depth.

The principal influence of turbidity on a diver is the effect on underwater visibility. It is almost useless to try to collect if underwater visibility is less than a fcot (30.5cm). Even visibilities less than 5 feet (1.5m) can impose serious limitations on activities such as photography or observing animals that swim rapidly.

WATER MOVEMENTS

Many agents cause water movements. Waves, wind, and tides are among the most important influences. Movement of short duration (such as that produced by waves) is called *surge*; longer-lasting movements (such as those generated by tides and wind) are called *currents*. Many organisms are sensitive to the intensity of surge and currents; some require constant movement, others need very calm water, while many flourish under intermediate conditions. A good rule of thumb for the open coast is that the farther north one goes, the rougher the sea becomes. Offshore islands have a calming influence on those parts of the coast falling within the "wave shadow."

The major surface-current systems off California vary with location and season. The predominant flow is a large sluggish movement called the California Current. The California Current brings cold water south from the Gulf of Alaska. In autumn the California Current is close to the coasts of northern and central California, but a hundred or so miles (161km) from shore off southern California. The coast turns easterly at Point Conception to form an indentation called the Southern California Bight (see Fig. 5), but the California Current continues southward at Point Conception and does not flow eastward into the Bight. The Current finally trends easterly off Baja California, reaching the coast at Punta Colnett, about 120 miles (193km) south of our Mexican border. A portion of the flow turns north here and moves up into the Southern California Bight.

The California Current flows close to the shores of northern and central California from about August to November. The nearshore flow reverses from time to time between November and April. The northward movement, known as the Davidson Current, usually weakens after April. The California Current, however, still remains offshore because of strong winds at this time from the north and northwest. The winds push the surface waters seaward. Replacement comes from deeper water–a process called upwelling, sometimes known as the Oceanic Period, is prominent from April to August in northern and central California.

UPWELLING

Vertical movement of water, called *upwelling*, occurs in certain areas of the coast. Although upwelling currents are slow compared to horizontal currents, the upward motion is very important because it brings cold, nutrient-rich deep water up near the surface. When marine organisms die, they usually sink to the bottom. Bottom waters thus accumulate the nutrient minerals resulting from decomposition, but surface waters become impoverished by the process. Upwelling restores nutrients to the brightly illuminated surface waters so that the minerals can be incorporated into growing plants.

Upwelling occurs when the wind moves surface water in an offshore direction. The deficit is filled by water rising from below (Fig. 2). Because of frictional drag in liquids, a wind-generated current flows at a 45° angle to the right of the wind direction in the Northern Hemisphere (it flows to the left in the Southern Hemisphere). Prevailing winds along most of California blow from the north or from the west. Such winds will move water away from shore where the coast tends to be oriented in an east-west direction. This orientation occurs on

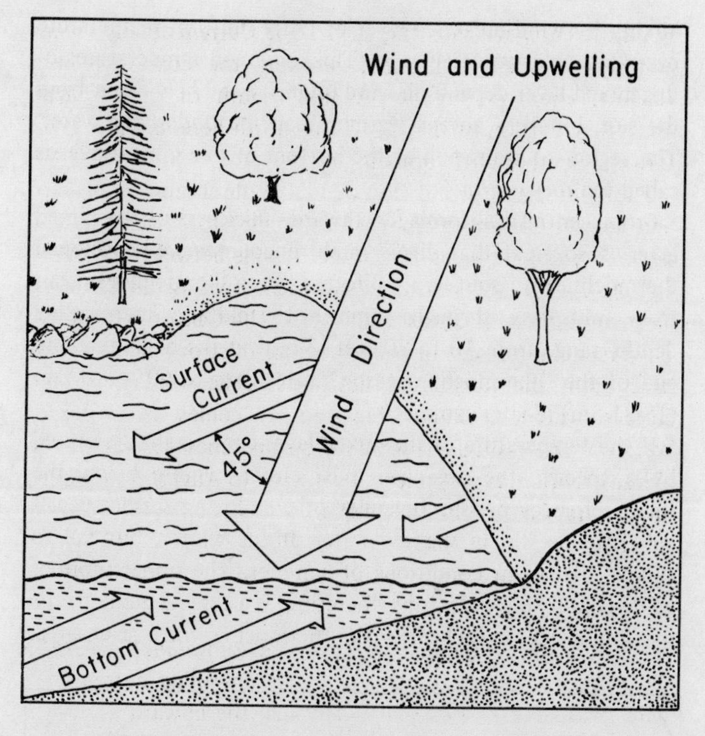

Fig. 2. Diagram showing 45° orientation between surface current and the wind that generates it. Along those portions of the coast where surface current is directed offshore, bottom currents are induced toward shore, causing upwelling.

the south side of headlands and along the north shore of bays and inlets, and such locations tend to have colder water than adjacent areas oriented otherwise. The cold regions are typically rich in life because of the presence of upwelling.

THERMOCLINE

Besides generating waves and current, wind also mixes the upper layers of the sea. In the absence of wind, water temperature would decrease uniformly from top to bottom (the surface is always warmest because of heating by the sun).

Mixing by wind creates a layer of fairly uniform temperature overlying cold water (Fig. 3). Thickness and temperatures of this mixed layer depend on wind intensity and heat input from the sun, but it is always warmer than the underlying water. The region of transition at the bottom of this mixed layer is called the *thermocline*.

From central California northward, thickness of the mixed layer is so great that divers rarely encounter well-developed thermoclines. In southern California, sharp thermoclines occur from midspring through summer to midfall. Thermocline depths range from 10 to 20 feet (3-6m) at the beginning and end of the "thermocline season" to as deep as 60 to 70 feet (18-21.5m) in later summer.

If the temperature of the mixed layer exceeds that tolerated by a species, the organism must either migrate below the thermocline or perish. Juveniles of cold-loving sessile species will often settle in shallow water during winter, but fail to survive the warm conditions of summer. The upper limit of *adult* cold-loving organisms, therefore, is a general guide to the maximum depth reached by the thermocline for a given area.

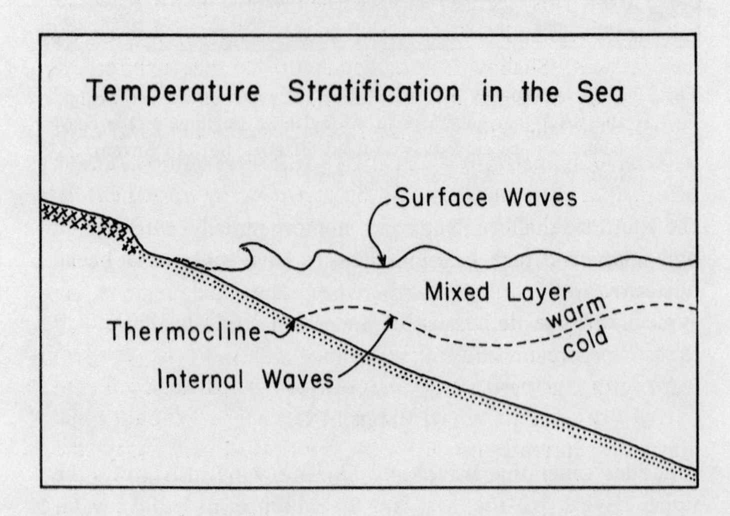

Fig. 3. Cross-sectional view of the nearshore region, showing thermocline and the surface and internal waves.

CHARACTERISTICS OF THE BOTTOM

SHELF WIDTH

Along most of the California coast the bottom slopes gently and fairly uniformly out to the edge of the continental shelf. There is substantial variation in slope angles from one region to another, however, so that shelf width is variable. North of San Francisco the shelf is 10 to 20 miles (16-32km) wide, but usually extends less than half this distance south of Monterey. Our chief concern is with the shallow littoral zone which usually lies within two miles of the beach. Wide shoal areas extend off Point St. George, Crescent City, San Francisco, Ventura, Los Angeles, and San Diego.

ROCK AND SEDIMENT RELATIONSHIPS

The different types of bottom and the influences that they exert on associated communities are discussed in detail in Chapter 4. This section is concerned with certain relations between rock and sediment that are peculiar to the marine environment. Shallow marine sediments do not exhibit the permanency of sands and soils in the terrestrial environment. On dry land the sediments are relatively immobile, and most plants and animals can spend their entire lives without undue hazard that their substrate will be destroyed by movement. In the sea the shallow sediments are continually shifting; an organism may face danger of excavation today and burial tomorrow.

Sediments are derived from land erosion and are carried to the sea by streams and rivers. Sediment transport occurs when waves bring the particles into suspension; the particles are then moved by current to a different location. In California, prevailing currents usually flow southward and carry the sediments slowly along as a massive river of sand; at certain areas or times, however, the flow is opposite. Variations in sediment volume and rates of transport are caused by changes in rainfall, current velocities, and wave characteristics. Flood control construction as well as construction of groins and jetties can profoundly alter sediment transport.

Organisms that inhabit sediments are usually able to cope with the transport of their substrate by a variety of adaptations (see Chapter 4). Problems arise, however, where sediment and rock environments meet. Rocky-bottom organisms are usually poorly adapted for life in a sedimentary environment, and sedimentary organisms often cannot survive on bare rock. When sand levels change substantially, rocky creatures can become buried or sand-dwellers can be left stranded on rocks, and at such times considerable mortality occurs. Places where these small-scale catastrophes happen frequently are typically rather barren or support only juveniles or short-lived organisms that have developed since the last major change. The burial and exposure problem arises rather frequently, particularly in rocky areas with considerable relief where sediments can accumulate in basins and depressions. Over certain rock bottoms where terrestrial erosion does not contribute much sediment (such as around rocky islands), coarse sands of biological origin can accumulate. These sediments are composed of fragments of shell and skeletons from mollusks, echinoderms, bryozoans, etc., and periodically can occur in sufficient amounts to bury unfortunate rock-dwellers.

CANYONS

The flow of sediments along the continental shelf either spills over the shelf edge or encounters one of the many great chasms that transect the shelf. These *submarine canyons* divert the sediments seaward in flows known as *turbidity currents* (the particles become suspended in the water, creating a fluid of heavy density). The shoreward ends of the canyons are usually steep gorges and may represent valleys eroded by streams during periods when sea level stood lower. Erosion has been continued by the turbidity currents that occasionally sweep down the canyon axis with the force of an avalanche. Canyons also collect seaweed and other organic debris that is moving along the shelf. At times this drift is so plentiful that the floor of the canyon looks like a garbage dump. Nonetheless, the material has nourishment value and supports a host of scavengers and decomposers. Biological aspects of submarine canyons are discussed in Chapter 4.

ZONATION

Subtidal zonation patterns exist, but are usually not as sharply defined as in the intertidal. Along open coasts there is typically a transition at depths of 15 to 25 feet (4.5-7.5m) (Fig. 4). Organisms above this demarcation must cope with the violence of wave surge, so rather sturdy forms are the rule. More delicate species do not appear until the greater depths have moderated wave action. In protected areas, however, fragile forms often occur in quite shallow water. In southern California, all shallow zones are exposed to warm water in summertime and are thus colonized primarily by species tolerant of elevated temperatures.

At depths of 30 feet (9m), the delicately structured species become more abundant and a scattering of cold-loving forms appears for the first time, in an area where surface layers are warmed seasonally. The next transition usually begins at a

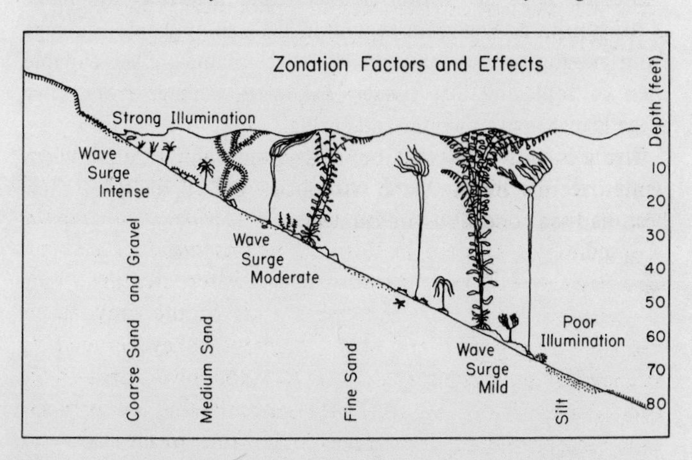

Fig. 4. Cross-sectional view of the nearshore region, showing zonal distributions of wage surge, illumination, and sediment size.

level where illumination no longer supports abundant attached vegetation. The depth here depends on average turbidity and generally lies between 50 and 100 feet (15-31m) in central and southern California; it may be as shallow as 10 to 20 feet (3-6m) in northern California. At greater depths grazing animals become scarce and are replaced by scavengers, detritus collectors, and filter feeders.

SEASONAL CHANGES

A number of seasonal changes occur in California waters. Wave surge becomes intense in winter and spring; this kicks up sediment from the bottom and reduces underwater visibility. Swollen by winter rains, rivers discharge more nutrient minerals, and microscopic suspended algae, the phytoplankton, quickly increase, often discoloring water by midspring. Again underwater visibility and turbidity are affected.

Intense wave surge enhances sediment transport. In general, beach erosion occurs in winter and sediments are moved offshore; the processes reverse in summer and beaches build up at the expense of deposits beyond the surf line.

Shorter days in winter reduce time available for plant photosynthesis. Seaweeds in winter often show greater damage from grazing animals; presumably these plants are less capable then of replacing lost tissues. Likewise, damage from wave surge is maximal in winter and spring.

Diving is most pleasant from midsummer until late fall when winter storms begin. Water visibilities are typically at their best, and sea conditions are calmest.

Chapter 3

MAJOR COASTAL SUBDIVISIONS

The California mainland (Fig. 5) is conveniently divided into three geographic regions: *northern* (Oregon border to San Francisco); *central* (San Francisco to Point Conception); and *southern* (Point Conception to the Mexican border). The marine fauna and flora of northern and central California are quite similar except at the extreme limits. Above Cape Mendocino certain subarctic species become more abundant, and south of Morro Bay a scattering of warm-water species appears. Southern California has a distinctive complement of warm-water organisms generally absent from the more northern subdivisions.

The northern portion of Baja California is subjected to intense upwelling in certain locations. It displays heterogeneous assemblages of both cold- and warm-water species, and is sufficiently distinct to justify classification as a major subdivision separate from southern California.

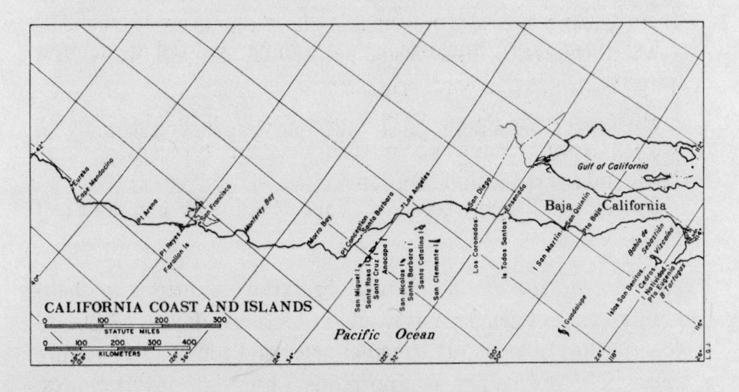

Fig. 5. Chart of the coast of California and northern Baja California and the offshore islands.

Finally, the large islands off California and Baja California contain unique environments and comprise the last of our major subdivisions. Water clarity here is nearly always much better than that found in mainland environments. Exposure to the sea ranges from extreme protection (without the siltation problems characteristic of mainland protected areas) to subjection to the most violent water motion. Island marine life usually prefers about the same water temperature suitable for nearby mainland creatures. Thus the Farallones (off San Francisco) support strictly cold-water species, while the lee shore of Cedros Island (Baja California) displays tropical forms.

NORTHERN CALIFORNIA–OREGON BORDER TO SAN FRANCISCO

GENERAL CONDITIONS

Water Temperatures: Winter: 46° to 50°F (7.77°-10°C). Summer: 52° to 58°F (11.11°-14.44°C).

Thermocline: Poorly developed or totally absent at shallow depths.

Exposure: Very rough seas; late summer and fall are calmest seasons.

Turbidity: Generally poor, but can improve seasonally in certain places.

Upwelling: Present south of Cape Mendocino and south of Point Arena.

Comments on Biota: Marine vegetation luxurious in sheltered areas, but generally restricted to shallow depths (20 to 30 feet [6-9m]). Areas off irregular coast and rocky islets usually very rich, displaying an enormous variety of invertebrates. Kelp beds largely formed of Bull Kelp (*Nereocystis*). Giant Kelp (*Macrocystis*) occurs principally between Fort Bragg and Point Arena.

Exploitation: Underwater largely free from influence of man. Access to beaches often difficult because of trespass problems. Many excellent public parks and beaches and commercial landings provide principal access.

Affinities: Fauna and flora are principally cold-temperate.

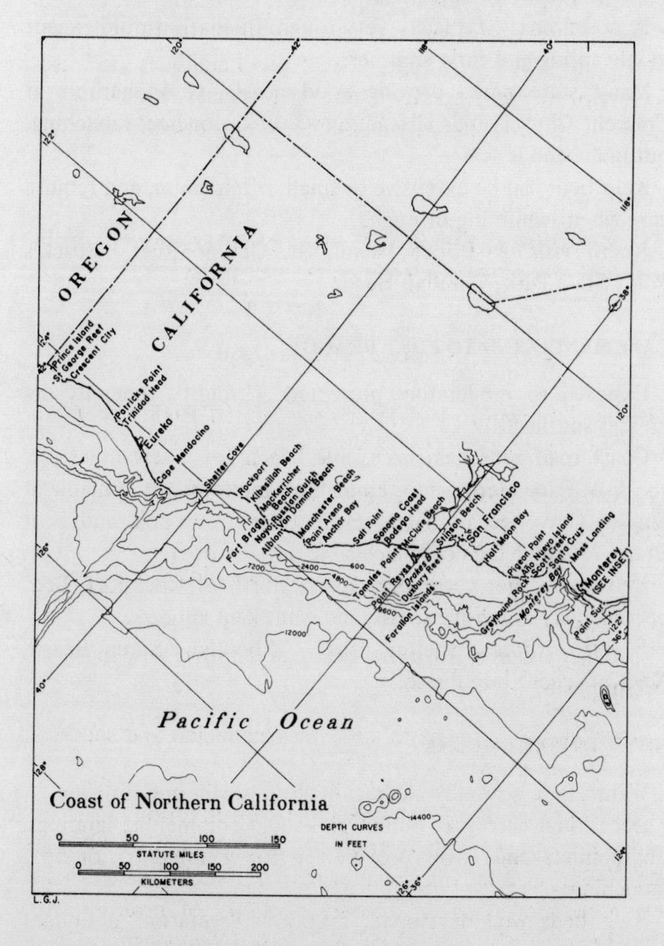

Fig. 6. Chart of northern California and a portion of central California, showing locations of diving sites discussed in the text. (See Fig. 20 for Monterey inset.)

Around Cape Mendocino a few subarctic species begin to appear, becoming somewhat commoner farther north.

REGIONAL DETAILS

OREGON BORDER TO EUREKA

Exposed coast, typically very rough. Intensely turbid except in late spring and early summer.

Many state parks provide good access. Good harbors at Crescent City, Trinidad Head, and Eureka for boat launching, but boat rentals scarce.

Kelp beds can be extensive or small; often dense, and require care when swimming through.

Worth visiting: Prince Island, St. George Reef, Patrick's Point State Park, Trinidad Head.

CAPE MENDOCINO TO FORT BRAGG

Exposed to moderately protected. Turbidity generally decreases southward.

Coast road runs near accessible beach for a few miles just south of Cape Mendocino. Launching and some lee available at Shelter Cove. Occasional access between Rockport and Fort Bragg. Much of the coast accessible only by boat.

Kelp and other vegetation sparse north of Rockport. Beds become very extensive south, and giant kelp appears.

Worth visiting: Shelter Cove, Kibesillah State Beach, MacKerricher State Beach.

NOYO TO POINT ARENA

Waters here generally clearest in all of northern California.

Occasional access at public parks and commercial landings. Many inlets and coves provide lee for water entry or boat launching.

Kelp beds well developed, luxurious vegetation, abundant fauna.

Worth visiting: Russian Gulch State Park, Mendocino (especially off Heeser Drive), Van Damme State Park, Albion, Manchester Beach State Park, Point Arena.

POINT ARENA TO BODEGA BAY

Generally exposed coast, water clarity variable from poor to moderate.

Access generally difficult except from Russian River to Bodega. Some access available over private land by paying trespass fee. A state underwater park is located at Salt Point.

Luxuriant vegetation in protected areas, rich fauna, kelp beds poorly developed.

Worth visiting: Anchor Bay, Salt Point State Underwater Park, Sonoma Coast Beaches State Park, Bodega Head and Bodega Bay.

BODEGA BAY TO SAN FRANCISCO

Major headlands provide some lee, but very rough seas occur in exposed areas. Water often very turbid.

Access often limited by ruggedness of terrain, as well as private ownership. A scattering of public beaches are principal areas for entry. Boats available at Tomales Bay.

Traces of kelp.

Worth visiting: Tomales Point, McClure Beach, Drakes Bay, Duxbury Reef, Stinson Beach.

CENTRAL CALIFORNIA–SAN FRANCISCO TO POINT CONCEPTION

GENERAL CONDITIONS

Water Temperatures: Winter: 48° to 53°F (8.88°-11.67°C). Summer: 52° to 64° (11.11°-17.78°C) on open coasts.

Thermocline: Usually absent or too deep and poorly developed to be of importance to divers.

Exposure: Rough seas occur frequently, particularly in winter and spring.

Turbidity: Variable from extremely clear to poor. Best clarity in late summer and fall.

Upwelling: Probably occurs south of headlands, but lack of a thermocline inhibits appearance of clearly defined cold-water regions.

Comments on Biota: Sea Otter herds present from Monterey to Port San Luis. Marine vegetation often luxurious. Kelp beds may be mixtures of *Nereocystis* (Bull Kelp) and *Macrocystis* (Giant Kelp); extensive beds at Pacific Grove, near Big Sur, San Simeon, and Point Arguello.

Exploitation: Influence of man variable. Greatest effects near ports such as San Francisco, Monterey, and Morro Bay. Almost virginal coastline from Point Lobos to Point Estero and around Point Buchon, but usually accessible only by boat. Region north of Point Arguello closed to public because of Vandenberg Air Force Base. Intensive commercial diving for abalone south of Monterey.

Affinities: Fauna and flora are primarily cold-temperate. From Morro Bay south, a few warm-temperate species appear.

REGIONAL DETAILS

SAN FRANCISCO TO DAVENPORT

Exposed coast–often rough and turbid.

Several excellent state parks provide good access in San Mateo County. Harbor at Half Moon Bay.

Kelp beds generally small and of fringing type.

Worth visiting: Pigeon Point, Año Nuevo Island, Scott Creek, Greyhound Rock.

MONTEREY BAY

Slightly more protected, but can be rough and turbid.

Good port facilities at Santa Cruz, Moss Landing, Monterey Harbor.

Kelp beds small, except near Santa Cruz.

Worth visiting: Natural Bridges State Beach, Monterey Coast Guard jetty, Del Monte kelp bed.

PACIFIC GROVE TO POINT LOBOS

Pacific Grove protected, can be turbid. South of Point Piños exposed, clarity usually good.

Access generally good. Permission required to dive in Point Lobos State Reserve.

Large kelp bed at Pacific Grove, fringing beds elsewhere.

Worth visiting: Fine diving almost anywhere. Carmel Canyon outstanding and accessible from the beach.

BIG SUR COASTLINE

Exposed coast–often rough, but usually clear because bottom is predominantly rock.

Access poor by land–generally requires a boat for diving.

Large offshore kelp beds; many islets and pinnacles; dropoff at shoreline usually steep.

Worth visiting: Interesting diving throughout, human influence negligible; State Underwater Park near Big Sur.

SAN SIMEON COASTLINE

Exposed to semi-protected. Often rough, but clarity usually good.

Beach access good near San Simeon; few boat launching facilities.

Extensive kelp beds, particularly in semi-protected areas. Offshore islets.

Worth visiting: San Simeon State Beach, Point Estero.

MORRO BAY TO POINT CONCEPTION

Exposed to completely protected coast; turbidity variable.

Good port facilities at Morro Bay and Port San Luis. State parks and public beaches provide access. No public access between Point Sal State Beach and Jalama.

Fringing kelp beds frequent, large bed south of Point Arguello. Islets off Point Buchon.

Worth visiting: Morro Rock, Lion Rock, Point Conception area.

SOUTHERN CALIFORNIA–POINT CONCEPTION TO MEXICAN BORDER

GENERAL CONDITIONS

Water Temperatures: Winter: 50° to 54°F (10.00°-12.22°C). Summer: 64° to 74°F (17.78°-23.33°C) at surface, 50° to 65°F (10°-18.34°C) below thermocline. Warmest season is August to mid-September.

Thermocline: Usually absent or poorly developed in winter. Noticeable about April at depths of about 30 feet (9m). Continues until about October, sometimes descending as deep as 50 to 70 feet (15-21.5m) during warm spells.

Exposure: Offshore islands and orientation of coast offer protection from groundswell originating from north to west. South of Santa Barbara, strong southern groundswell often present during summer. Local storms can generate strong local swell in the region between the mainland and the islands, but such disturbances quickly subside after the wind drops.

Turbidity: Variable seasonally and from one place to another. Plankton causes poor conditions throughout coastal regions from about March to mid-August, occasionally into September. Visibilities usually low following heavy rains, clearing up first near rocky headlands.

Upwelling: Well developed wherever coast trends in an east-west direction (e.g., Point Conception to Santa Barbara). Likewise found south and southwest of major headlands (e.g., Point Dume, Palos Verdes, Dana Point, Point Loma). Major kelp areas often coincide with upwelling regions.

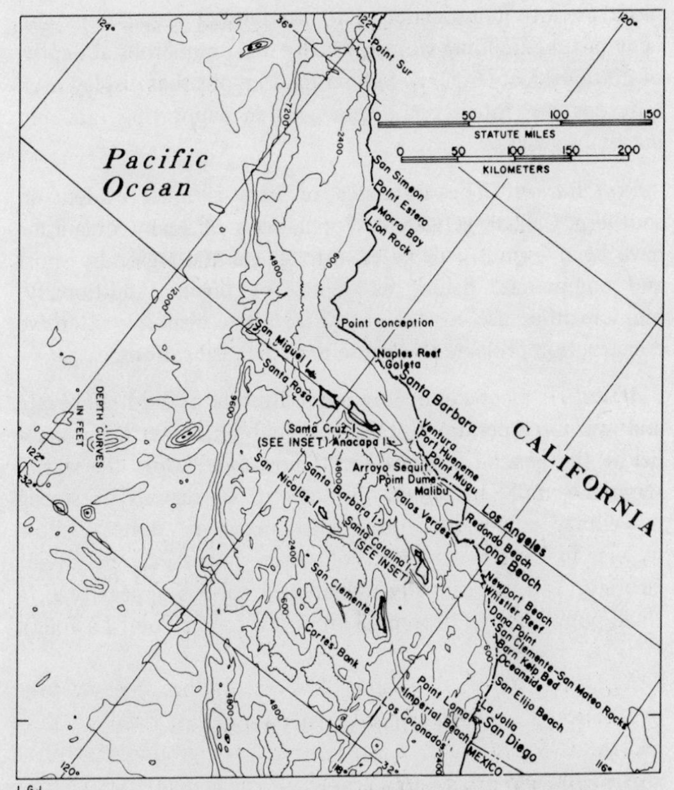

Fig. 7. Chart of southern California and a portion of central California, showing locations of diving sites discussed in the text. (See Figs. 9a and 9b for island insets).

Comments on Biota: In general the richest areas are the kelp beds, and these occur frequently, so that the habitat is usually available throughout southern California. In places swarms of sea urchins dominate the bottom (possibly because of extermination of their principal predator, the Sea Otter), and such areas tend to be barren. On sand bottom, rich and interesting biota are often associated with sand-dollar populations and beds of tube-building worms, usually most numerous at depths of 20 to 40 feet (6-12m). Submarine canyons that display rock outcrops are interesting regions, often supporting rare and unusual species.

Exploitation: The influence of man is most evident in southern California waters. Populations of many organisms have been seriously depleted throughout the region by sport and commercial fishing as well as by divers. Additionally, municipalities use coastal waters for waste disposal; extensive construction projects also cause periodic dislocations.

Affinities: Fauna and flora are a mixture of cold-temperate and warm-temperate. Usually the cold-temperate occur only below the general level of the thermocline, while the warm-temperate may be found throughout. In regions of strong upwelling, cold-water forms may occur in quite shallow waters. In bays and bights, cold-water forms may be absent entirely, and occasionally subtropical species appear (e.g., a small population of broomtail grouper inhabits Point La Jolla).

REGIONAL DETAILS

POINT CONCEPTION TO VENTURA

Generally well-protected, some exposure near Point Conception. Turbidity often extensive, probably because of plankton induced by upwelling conditions. Medium-sized sewer outfall at Santa Barbara contributes locally to turbidity.

Frequent state parks provide beach access from Gaviota eastward. Good boat launching facilities and rentals at Santa Barbara.

Extensive kelp beds—those on sandy bottoms tend to be less rich in species than those on rocky bottoms.

Worth visiting: Point Conception, Naples Reef, reef at Goleta.

VENTURA TO POINT DUME

Generally well-protected, often turbid.

Coast highway and public beaches furnish good access. Boat launching at Ventura and Port Hueneme.

Best kelp beds between Point Mugu and Zuma Beach, irregular and interesting bottom. Submarine canyons near Point Mugu and Point Dume.

Worth visiting: Sand dollar bed off Ormond Beach, kelp bed off Sequit canyon, shallow reefs at Point Dume.

POINT DUME TO REDONDO BEACH

Well-protected, fairly turbid except Point Dume to Malibu can be clear seasonally.

Access from public beaches and coastal highway. Boat launching at Paradise Cove, Malibu, Marina Del Rey, Redondo Beach.

Good kelp beds from Point Dume to Malibu. Submarine canyon at Redondo Beach.

Worth visiting: Reefs at Paradise Cove, kelp bed at Corral Canyon.

PALOS VERDES TO NEWPORT BEACH

Exposed to local swells and some groundswells. Usually turbid from waste disposal and local construction projects. Northwest Palos Verdes areas clear seasonally.

Access from public beaches and harbors. Boat launching at Los Angeles Harbor, Long Beach Harbor, Alamitos Bay, Newport Bay.

Former kelp beds at Palos Verdes will probably be restored during next ten years. Submarine canyons off Point Fermin and Newport Beach.

Worth visiting: *Dominator* wreck at Rocky Point, Point Vicente, west Newport jetty.

NEWPORT BEACH TO OCEANSIDE

Exposed to local swells but protected to north and west from groundswell. Some seasonal turbidity.

Access from public beaches, some private beaches, coastal highway, and harbors. Boat launching at Dana Point Harbor, San Clemente pier, Oceanside Harbor.

Kelp beds fairly frequent from Newport Beach to San Onofre. Occasional islets.

Worth visiting: Whistler and Pequegnat's reefs, Heisler Park, Dana Point kelp bed, Barn Kelp bed.

OCEANSIDE TO POINT LA JOLLA

Some protection from north groundswell, exposed to west and southern groundswell and local swell. Turbidity seasonal, but Point La Jolla usually clearest area on coast.

Access from public beaches and coastal highway. Boats can be launched across the beach in areas where surf is low (e.g., at head of La Jolla submarine canyon).

Kelp beds at frequent intervals along the coast. Large submarine canyon at La Jolla, accessible from the beach.

Worth visiting: Kelp bed at San Elijo State Beach, La Jolla submarine canyon, Point La Jolla. La Jolla Bay is now an underwater park operated by the City of San Diego.

POINT LA JOLLA TO MEXICAN BORDER

Generally exposed to all but direct northern swell. Turbidity mostly seasonal, but water clarity can be poor at almost any time near entrances to Mission Bay and San Diego Bay.

Access from public beaches and harbors. Boat launching at Mission Bay, San Diego Bay.

Kelp beds at La Jolla and Point Loma; former bed at Imperial Beach may eventually be restored.

Worth visiting: La Jolla kelp bed, Point Loma sea cliff.

NORTHERN BAJA CALIFORNIA

GENERAL CONDITIONS

Water Temperatures: Winter: 50° to 56°F (10°-13.33°C). Summer: 56° to 75° (13.33°-23.89°C) at surface, 48° to 64°F (8.88°-17.78°C) below thermocline. Warmest season is August to September.

Thermocline: Absent or poorly developed in winter. First noticeable in midspring in open coasts and bays. Appears later in upwelling areas. Generally well developed by July. Subsides in midautumn.

Exposure: Except in the lee of islands and headlands, the coast is exposed to rough seas and is more reminiscent of central California than of southern California. Strong winds may appear suddenly and without warning, particularly in the southern part of this region, generating extremely choppy seas in a matter of minutes. Craft with little freeboard should not be operated far from shore.

Turbidity: Somewhat seasonal, although generally fairly clear. Plankton blooms, storms, and runoff from torrential rains can affect water clarity from January through late spring.

Upwelling: More noticeable here than farther north, possibly because the cold upwelling areas contrast more sharply with the warm bights and bays. Nonetheless, many areas are intensely cold well into summer, perhaps reflecting persistence of wind regimes.

Comments on Biota: Northern Baja California gives a general impression of extreme species diversity. Certainly a very wide variety of contrasting environments occur in close proximity to each other, and this perhaps offers an opportunity for intermingling of species that would not ordinarily occur together. The coast is dotted with kelp beds ranging from fringing stands to vast stretches of submarine forest.

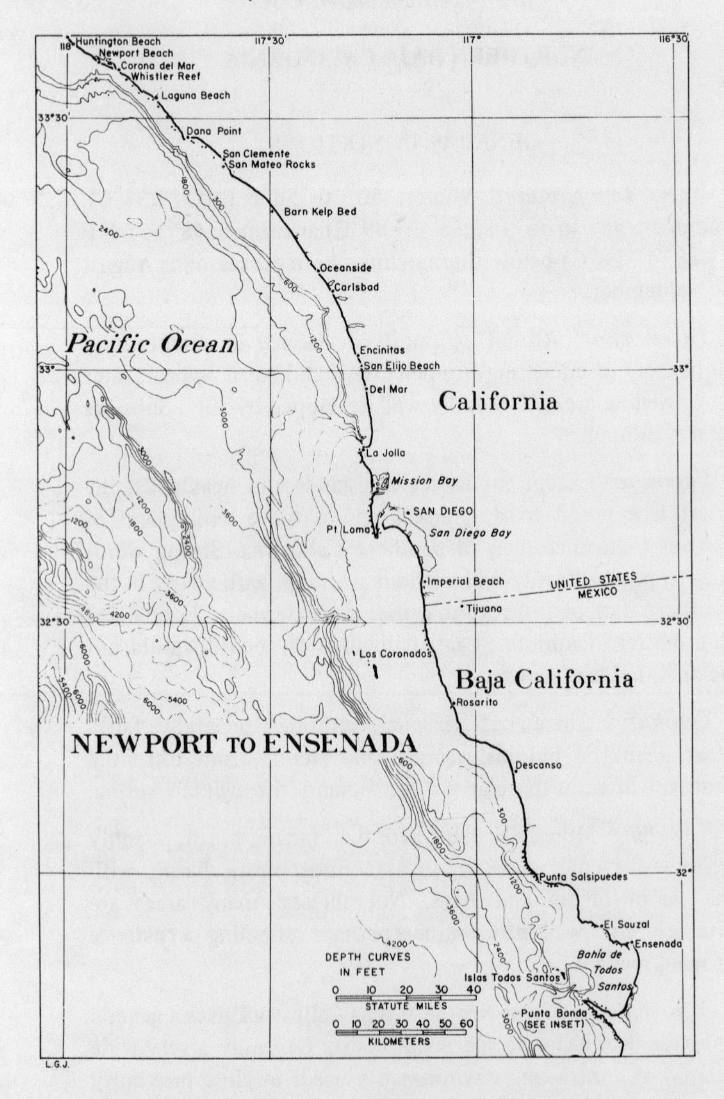

Fig. 8. Chart of region from Newport Beach, California, to Punta Banda, Baja California, showing locations of diving sites discussed in the text. (See Figs. 15 and 16 for insets.)

Exploitation: Human influence is greatest in the regions nearest the border, decreasing markedly a few miles south of Ensenada. Nowhere does it reach the proportions commonly seen in southern California.

Affinities: Affinities of northern Baja California are broad, probably because of the variety of environments present. Cold-water species occur in the vicinity of Ensenada that are characteristic of central California but rare or absent throughout southern California. On the other hand, the warm-temperate fauna and flora of southern California are also abundant in northern Baja California. South of San Quintín Bay, subtropical forms appear in increasing abundance, while the southern part of Vizcaíno Bay becomes almost entirely subtropical.

REGIONAL DETAILS

U.S. BORDER TO PUNTA SALSIPUEDES

Exposed coastline, but good for diving because long stretches of calm weather occur in summer and fall. Turbidity seasonal, little influences from human activities.

Access generally good except where rugged terrain interferes. Some coastal areas privately held, but arrangements for access are usually easily negotiated locally. Boats must be launched through the surf.

Large offshore kelp beds, some islets.

Worth visiting: All areas are interesting.

ENSENADA AND VICINITY

Exposed and protected coastline. Turbidity seasonal and highest within Todos Santos Bay. Water clarity nearly always good at tip and along south side of Punta Banda.

Access generally good. Boat launching facilities in Ensenada, and concrete ramps are available in many of the tourist camps along the shore.

Small kelp beds along outer portions of Todos Santos Bay, interesting marshland in southeast region of bay. Islets and pinnacles along Punta Banda.

Worth visiting: Punta Banda tip, Papalote Bay (La Bufadora).

PUNTA SANTO TOMÁS TO SAN QUINTÍN BAY

Exposed coastline. Turbidity seasonal, water clearest off headlands.

Access good where roads to coast exist. Boat launching ramps at Puerto Santo Tomás and San Quintín Bay.

Large offshore kelp beds along much of coast. Occasional islets.

Worth visiting: South side of Punta Santo Tomás, Punta Piedras (site of *Tampico* wreck), San Ysidro Reef.

SOUTH OF SAN QUINTÍN BAY

Exposed coastline, water clarity generally fair to good.

Where coastal roads exist, beach access unimpeded. Boats launched through the surf. Coast generally accessible only by boat south of Rosario Bay.

Moderate kelp beds until Rosario; Rosario Bay and Geronimo Island sometimes support extensive kelp. Islets and pinnacles.

Worth visiting: Rosario Bay, Sacramento Reef.

THE ISLANDS

GENERAL CONDITIONS

Water Temperatures: Usually correspond fairly closely to nearest point on mainland, often being a degree or two cooler.

Thermocline: Corresponds to mainland circumstances for the particular region.

Exposure: Weather sides of islands usually exposed to rougher seas than the corresponding mainland, but lee sides typically well protected.

Turbidity: Generally good underwater visibility, occasionally affected by plankton blooms. Sheltered areas not affected by siltation, which is so common on the mainland.

Upwelling: Upwelling to be expected off southwest and southern shores.

Comments on Biota: Islands frequently support species absent or rare on the mainland. Likewise, certain plants and animals may flourish in special island environments well north or south of their usual ranges on the mainland. Bottom topography usually displays considerable relief.

Exploitation: Variable; human influence quite evident on tourist-frequented islands such as Catalina, Los Coronados. Such effects negligible on more remote islands.

Affinities: Most of the biota typically bear resemblance to the closest region on the mainland. San Miguel, Santa Rosa, and Santa Cruz islands support certain cold water organisms more characteristic of central California. On the other hand, Catalina and San Clemente islands contain primarily warm-temperate organisms with an admixture of subtropicals. Cedros Island displays warm-temperate communities on the west side and subtropical species on the east side (facing subtropical Vizcaíno Bay).

REGIONAL DETAILS

SAN MIGUEL AND SANTA ROSA ISLANDS

Best protection on southeast sides, water generally clear.

Privately owned, access only by substantial vessels that can face rough seas.

Fair-sized kelp beds. Sea Otters have been seen here in recent times.

Poorly explored underwater.

SANTA CRUZ AND ANACAPA ISLANDS

Fairly protected except when local storms in progress, generally clear water.

Privately owned, access by large vessels capable of 10- to 20-mile (16-32km) journeys.

Kelp beds, underwater caves.

Worth visiting: Gull Island, northeastern Anacapa.

SAN CLEMENTE AND SANTA CATALINA ISLANDS

Eastern sides protected, water clarity usually excellent.

San Clemente is government-owned, Santa Catalina privately owned. Access by boat. Military harbor at north end of San Clemente; Catalina harbors at Avalon, Isthmus Cove, Catalina Harbor, and others. Ferryboat and commercial air service to Catalina.

Fringing kelp beds most frequent; large beds on west side of San Clemente. Islets and pinnacles.

Worth visiting: The Hook, Farnsworth Bank, Ship Rock.

SAN NICOLAS AND SANTA BARBARA ISLANDS AND CORTES BANK

Rough weather frequent, protection only moderate even on lee sides of islands; water clear.

Government owned; San Nicolas operated as military base, Santa Barbara as game preserve. Access by large vessel only.

Kelp beds. Herds of Sea Lions and Elephant Seals.

Poorly explored underwater.

LOS CORONADOS ISLANDS AND ROCKPILE

Good lee on east sides, water usually but not invariably clear.

Government-owned (Mexico)–be sure proper boat papers and licenses have been obtained. Access by fairly small vessel, 18 ft. (5.5m) or longer.

Fringing to moderate-sized kelp beds. Sea Lions, a few Elephant Seals.

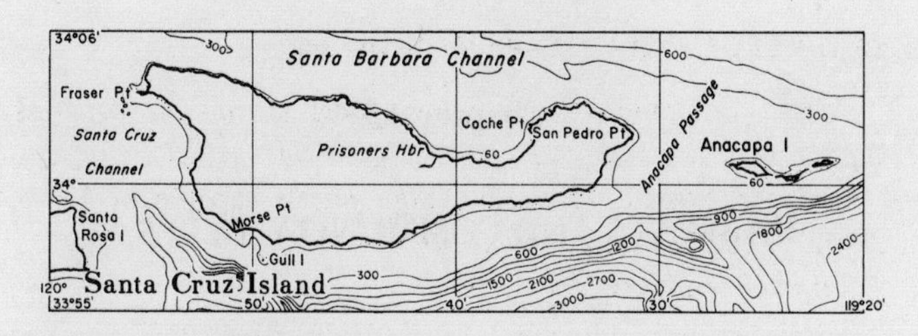

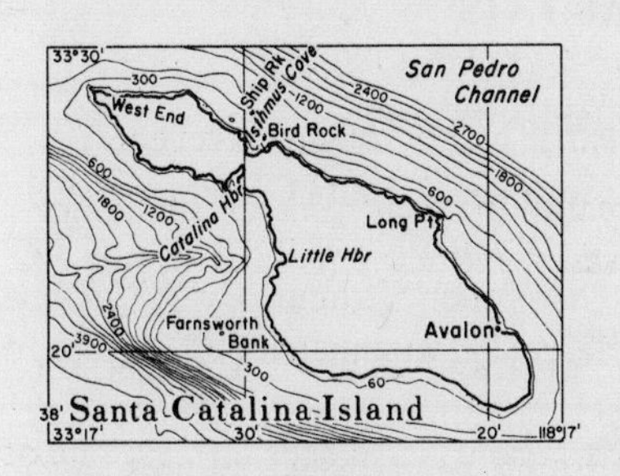

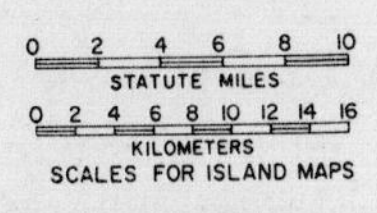

Fig. 9. a. Chart of Santa Cruz and Anacapa Islands. b. Chart of Santa Catalina Island.

Worth visiting: Middle Grounds, Jackass Rocks, Five-Minute kelp bed.

TODOS SANTOS, SAN MARTÍN, SAN GERONIMO, CEDROS, AND SAN BENITOS ISLANDS

Good lee available around Cedros, otherwise exposed; water clarity usually excellent.

Todos Santos Islands accessible by fairly small boat, otherwise large vessels only.

Moderate-sized kelp beds common. Whales, Sea Lions, Elephant Seals. Islets and pinnacles.

Worth visiting: Top-notch diving almost anywhere.

Chapter 4

HABITATS AND COMMUNITY TYPES

HABITATS

ROCKY OPEN COAST

Rocky bottoms vary greatly in their general submarine illumination, exposure to waves and currents, and degree of unevenness of the rocks.

Illumination is important primarily because it is needed for plant growth. It decreases as depth increases and is also affected by turbid waters issuing from rivers, harbors, and bays.

Wave surge also decreases as depth increases. Protection by headlands and offshore bars or islands diminishes the effects of waves and currents. Violent water movements favor some plant and animal species, but prevent others from becoming established. Intense wave surge also moves loose rocks and brings sand into suspension. Abrasion, caused by the moving rock and sand on stationary surfaces with which they collide, can destroy attached life and prevent recolonization.

The bottom topography of rocky areas varies from flat smooth surfaces, almost like pavement in streets, to highly convoluted formations creating cliffs, crags, pinnacles, and caves (Fig. 10). Loose rock is often superimposed, and occasionally occurs over thin sand deposits. Loose rock is classified according to size: *pebbles* represent a size range from coarse sand up to stones 1 or 2 inches (2.5-5.1cm) in diameter; *cobbles* are larger than pebbles, with diameters up to 10 or 12 inches (25.4-30.5cm) and *boulders* designates even larger loose rocks.

In general, the more irregular the rocks, the richer the communities associated with their surfaces. Flat pavement

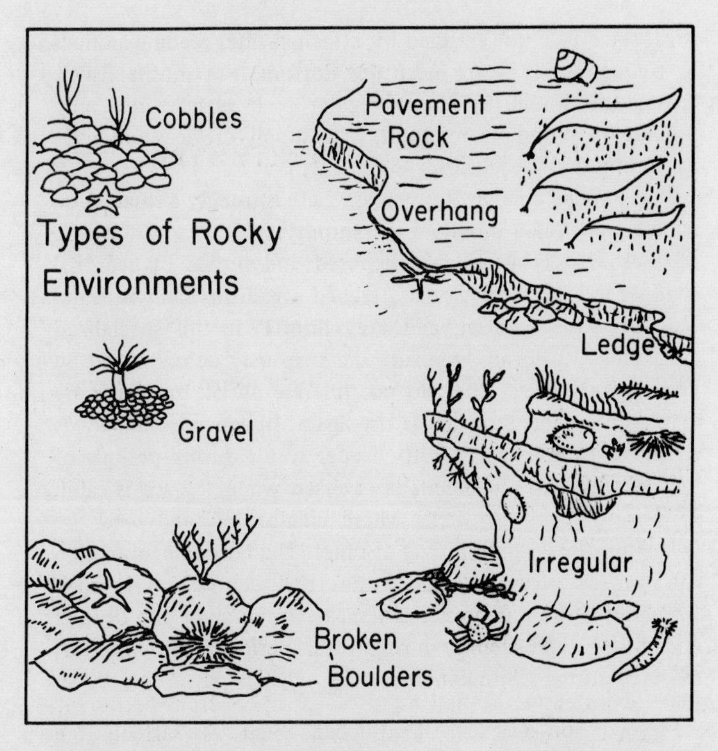

Fig. 10. Diagram showing different types of rocky-bottom environments.

rock is usually poorest, unless it is colonized by plants or sessile animals that convert the smooth surface to an irregular one. At the other end of the scale, submarine cliffs and pinnacles are often covered with a turf as much as 6 inches (15.2cm) thick with organisms piled on top of each other. These mats nourish clouds of fishes that hover in the water nearby. Small irregularities cause turbulence in water flowing over a rock surface, which apparently benefits sessile animals—perhaps by increasing food and oxygen supplies. Large irregularities generate even more turbulence. Very large irregularities may penetrate into the midwater currents, which are usually much stronger than bottom currents. The midwater region may be richer in suspended food particles and organisms, since it is not strained by attached filter-feeding animals as frequently as water near the bottom. For motile fishes, crustaceans, mollusks, etc., irregular rock surfaces not only provide an abundance of food, they supply crevices and holes for hiding from predators.

Where rocks emerge from sand, there is usually a small basal zone supporting a modified community. This zone is the strip of rock that is alternately exposed and buried by seasonal changes in sand level. Some attached organisms can withstand long periods of burial, and they flourish in this specialized environment. During exposure the strip may be colonized by juveniles of species that do not tolerate burial, but these are eliminated when sand levels rise again. In general, sand tends to move from beaches into deeper water during periods of large swells; the movement is reversed when the sea is calm. Low-lying rocks, or rocks where changes in sand level are great, may be entirely buried at times. The burial phenomenon can also be seen in small basins and depressions in rock formations; these depressions tend to accumulate bits of shell and other solid debris during calm periods, and lose this material during storm conditions.

Presence or absence of attached plant life affects the character of rocky-bottom communities profoundly. Many grazers such as urchins, abalone, kelp crabs, certain fishes, and a variety of snails depend on seaweeds for nourishment. Many other animals use attached plants for shelter or attachment. Holdfasts and stipe bundles of the larger kelps may offer protection from wave surge. The seaweed-dependent species become scarce or disappear as these plants become sparse (for example, as depth increases). Encrusting filter-feeding animals that strain the water for suspended food take over rock surfaces which, for one reason or another, cannot support a lush growth of plants. Lack of adequate light is the most common cause limiting plant growth. Thus filter feeders form thick turfs or mats beneath ledges and overhangs in shallow water. The heavily shaded bottoms below kelp beds and reefs deeper than 100 feet (30.5m) tend to be sparse in plants, but rich in such turfs.

SANDY OPEN COAST

Some sandy bottoms resemble terrestrial deserts. Here and there a fragment of seaweed or a worm tube breaks the monotony of the flat expanse. You can spend much time roaming across sandy wasteland in hopes of turning up an isolated treasure such as a lone halibut. In general, however, it is better to terminate a dive within a few minutes after descending into such areas.

Except for very calm shallow areas, seaweeds do not grow on sandy bottoms. Herbivorous animals must either consume drifting debris or filter their food as suspended particles in the water. When food imports are low, barren wastelands occur. The more interesting sandy communities are found where decomposing debris and detritus are abundant or where fairly continuous supplies of phytoplankton are available. The zones where various common animal communities may be found are usually fairly well defined (e.g., sand-dollar beds).

From the lower intertidal to depths of 20 to 40 feet (6-12m), a relatively rich fauna exists (Fig. 11). These animals are primarily nourished by the small plankton that are generally quite abundant in the upper layers of the sea. Below this rich layer, a zone of fairly impoverished bottom typically occurs, extending from depths of about 40 feet (12m) to roughly 100 to 130 feet (30.5-39.5m). Below this level, animals again become fairly abundant, probably nourished by settling debris.

The shallowest zone is an area subjected to varying sand levels as the seasons change. To survive here, animals must either be expert burrowers or be able to regain their buried condition rapidly after exposure, or in some way stabilize the substrate. Several clam species that are excellent diggers flourish within and just seaward of the surf line; most make good eating, and can be detected by paired siphons that may protrude slightly from the sand. Tube-building worms can stabilize sandy bottom, provided they occur in sufficient density. Thick clusters of sand dollars also occur in the shallow zone and probably contribute to substrate stability. Sometimes mixtures of animals—sea pens, burrowing anemones, sea

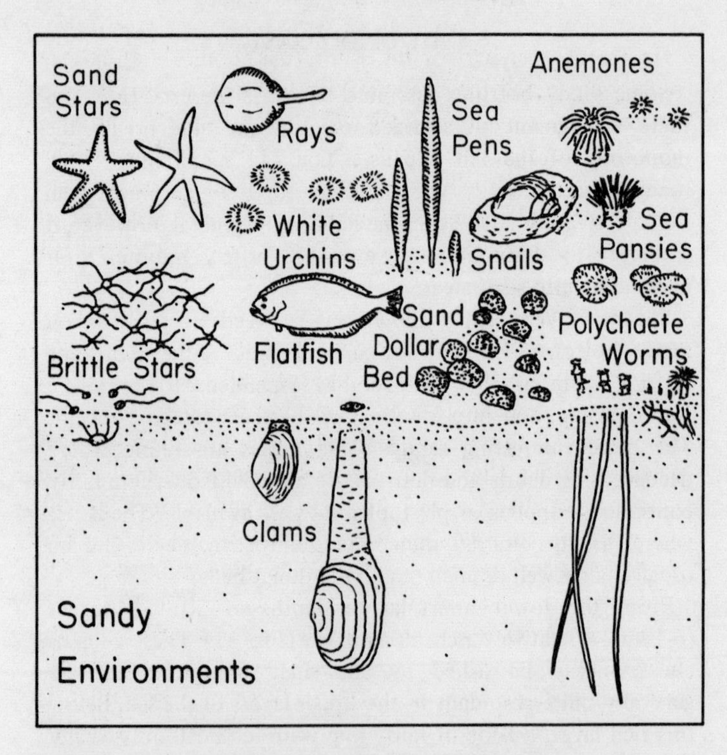

Fig. 11. Diagram of sandy environments and typical inhabitants.

pansies, worms, and mollusks–develop in sufficiently close association to stabilize their surroundings.

Changes in sand levels between depths of 40 and 130 feet (12-39.5m) are not as severe as in the shallow zone, but plankton concentrations are usually sparser. Most of the resident animals are burrowing worms, crustaceans, and mollusks, concealed in the top few inches of sand. Squid occasionally spawn in the barren zone. The adult animals mate close to the bottom; the female then buries the tip of the white pear-shaped egg-case in the sand. Both male and female die, and the bodies attract fishes, crustaceans, and marine mammals to the area. The eggs take several weeks to develop, and the egg-case may remain white or turn dark due to diatom encrustation.

In the deepest part of the scuba zone, numbers of animals lying on top of the sand frequently increase. Brittle stars become very common—either with the disk buried and only armtips showing, or completely exposed. True starfishes (Asteroids) are also fairly common, as are sea pens and the White Urchin (*Lytechinus*). Like the shallower zones, the first few inches of bottom is populated with a variety of worms and mollusks. Flatfishes are usually more common in this zone. Shallow depressions in the sand often contain a fascinating array of tiny shells, primarily remains of the burrowing fauna. Wave surge only stirs the bottom gently at these depths, and shifts in sand level are rarely a problem to the fauna.

Nutrition in the deepest zone probably depends greatly on settling detritus. There is nearly always a fine scum or dust-like sediment on the sea floor at these depths, formed by the accumulation of sedimenting microscopic organisms. The sands themselves are usually much finer than sediments of the shallow zones, and at times grade into muds.

Variations in this general pattern are common. Where sandy regions lie down-current from large beds of seaweed, the sediment-dwelling animals will have large quantities of drift available for nourishment. An environment near a bay will probably be exposed to higher-than-normal concentrations of microplankton at the shallow levels. Such variations affect both the species composition and the densities of the populations.

ESTUARIES AND BAYS

Strictly defined, an estuary is a region where salinity (saltiness) of the sea is affected by inflowing freshwater streams. Only specialized organisms that tolerate salinity changes survive in such an environment. Bays are fairly large bodies of seawater wherein wave action from the open sea is negligible because of protection from surrounding land. Bays may have estuarine portions where streams enter. Most southern California bays were true estuaries a few centuries ago when rainfall was apparently higher; consequently they are still colonized by typical estuarine organisms, since these species also tolerate full-strength seawater. During wetter winters these former estuaries are exposed to appreciable dilution. The change probably destroys any young non-estuarine species that may have developed and helps to maintain the estuarine character of the fauna and flora. Farther north the estuarine environment persists the entire year, due to continuous stream flow.

The floor of a typical bay or estuary ranges from fine sand to thick mud. Remains of dead plants and animals accumulate in sediments of the relatively stagnant regions. Intense decomposition creates a nutrient-rich black sand or mud lying a few inches below the sediment-water interface. In areas where substantial water flow occurs, the bottom is usually fine to coarse sand without underlying blackening.

In southern California and Baja California, shallow parts of the bays usually become much warmer than the open sea during summer. Organisms that do not tolerate higher temperatures do not survive. Bays farther north remain fairly cold the year round and can therefore sustain plant and animal species that occur in moderately deep waters of the nearby open sea. These cold-loving forms are unable to colonize the southern bays, although juveniles may appear during winter. This does not mean that the southern bays are necessarily species-poor. Many subtropical forms that reach their northern limits in southern California find ideal habitats in the bays. Open-sea species sometimes have a modified appearance when they occur in bays; many delicate forms can grow larger and develop more intricate structures in the quiet waters. Other species may be stunted—perhaps the bay is not an ideal environment for them, although they are able to survive.

One common sight in bays is an Eelgrass meadow (Fig. 12). Eel Grass (*Zostera*) extends upward 2 or 3 feet (61 or 91cm) from the bottom, and the meadow closely resembles tall terrestrial grasslands. Many fishes and invertebrates seek food and cover in the Eelgrass, and a host of animals burrow among the roots and feed on the decomposing material in the upper sedimentary layers. The typical eel-grass meadow is too thick

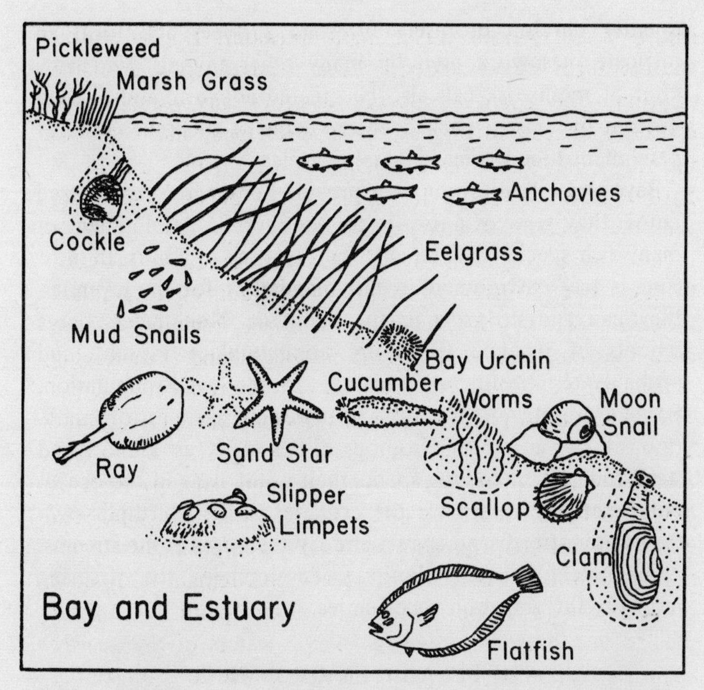

Fig. 12. Diagram of baylike environments, showing common fauna and flora.

for a human to swim through; either pass over the top or seek out the channels through the meadow where swifter current flows concentrate.

At depths greater than 10 feet (3m), solid surfaces in bays (rocks, empty shells, trash, etc.) are frequently densely colonized by animals. Deeper than 20 to 30 feet (6-9m) plants play only a minor role (the reverse of the usual situation in the open sea). Plants can, however, dominate when solid surfaces extend into shallow water and occupy substantial areas (jetties, ramps, long reefs, etc.). Perhaps a critical size is necessary for a plant colony to survive. Whatever the cause, the advantage is clearly with encrusting animals on small, dimly illuminated surfaces, and hard-shelled animals such as mussels, tube snails, or certain polychaete worms will frequently develop profusely on solid surfaces and form an intricate meshwork used by many other animals. The same kind of meshwork can also be created above a mud or sand bottom by closely packed encasements of tube anemones or parchment tube worms in dense colonies.

Bays and estuaries not only provide homes for a specialized fauna, they serve as nurseries for the juvenile development of many fish species that inhabit the open sea as adults. In most cases a bay environment is not mandatory for the juveniles, they can and do grow in the open sea. Nonetheless, large schools of juvenile fishes are commonplace in bays and probably contribute significantly to the total population. Besides the nursery function, estuaries are avenues for migration of important anadromous fishes such as salmon and steelhead. These species spend their adult lives in the ocean, but eventually return via the estuaries to their birthplaces in freshwater streams, to spawn and lay their eggs in the streams. After hatching and preliminary development the offspring return to the sea, again via estuaries.

THE PELAGIC ENVIRONMENT

Communities of organisms swimming or suspended in the water are fleeting–sometimes here today and gone tomorrow. Their movements are controlled by currents. Many variable forces produce ocean currents (wind, tides, swell, among others), so the net water movement is usually difficult to predict. If a pelagic community is spread over only a small area–an acre to several square miles–only a few minutes or hours will be needed for its passage by a fixed point if currents are strong. Alternatively, a large plankton bloom, extending along several hundred miles of coast, may persist for weeks or months if currents are weak.

The smaller plankton usually concentrate fairly close to the surface at depths from 10 to 100 feet (3-30.5m). Many larger plankton have substantial swimming abilities. They often descend hundreds of feet in daytime to avoid light, and then return to the surface at night to feed. Small plankton tend to be more concentrated in inshore waters, dispersing as distance from shore increases. Planktonic concentrations are usually dense above the thermocline, becoming much sparser below; but occasionally the reverse can occur, or there may be no difference whatever. Very dense concentrations of plankton may colloquially be called "tide" and are referred to by their color (*red tide, yellow tide*, etc.).

In the lists of marine animals in Chapter 8, we have omitted most of the plankton because identification usually requires microscopic examination and is beyond our scope. Some general remarks, however, are necessary and appropriate. Single-celled plants (*phytoplankton*) are usually the most abundant component of the microplankton. Two groups usually dominate: diatoms and dinoflagellates. *Diatoms* have external shells of silica that typically are sculptured with fine lines; this group tends to be most abundant in early spring. *Dinoflagellates* are usually enclosed by plates of cellulose, and are propelled by two hair-like flagella. Some contain chlorophyll (necessary for photosynthesis), but other species lack it and are considered animals. Dinoflagellates typically reach maximum concentrations in late spring. If the dominant species is *Gonyaulax polyedra* or *Noctiluca*, the water becomes coffee- to vermilion-colored and the familiar summertime red tide results. The next step in the pelagic food chain is grazing animals (*zooplankton*), usually larger than the phytoplankton but still microscopic. These range from larvae of benthic animals, only temporary members of the zooplankton, to permanent plankton such as tiny crustaceans and protozoans.

The microplankton are consumed by larger plankton including small pelagic fishes (sardine, anchovy, etc.) as well as many pelagic and benthic invertebrates (Fig. 13). These larger animals are known as *filter feeders* since they use tiny elaborate sieves, nets, mucus, and other mechanisms for straining plankton from the water. Higher links on the pelagic food chain involve the large fishes such as yellowtail, barracuda, and tuna. The top of the food chain culminates in large marine mammals such as porpoises and killer whales, as well as large sharks of the open sea. Curiously, the largest of all

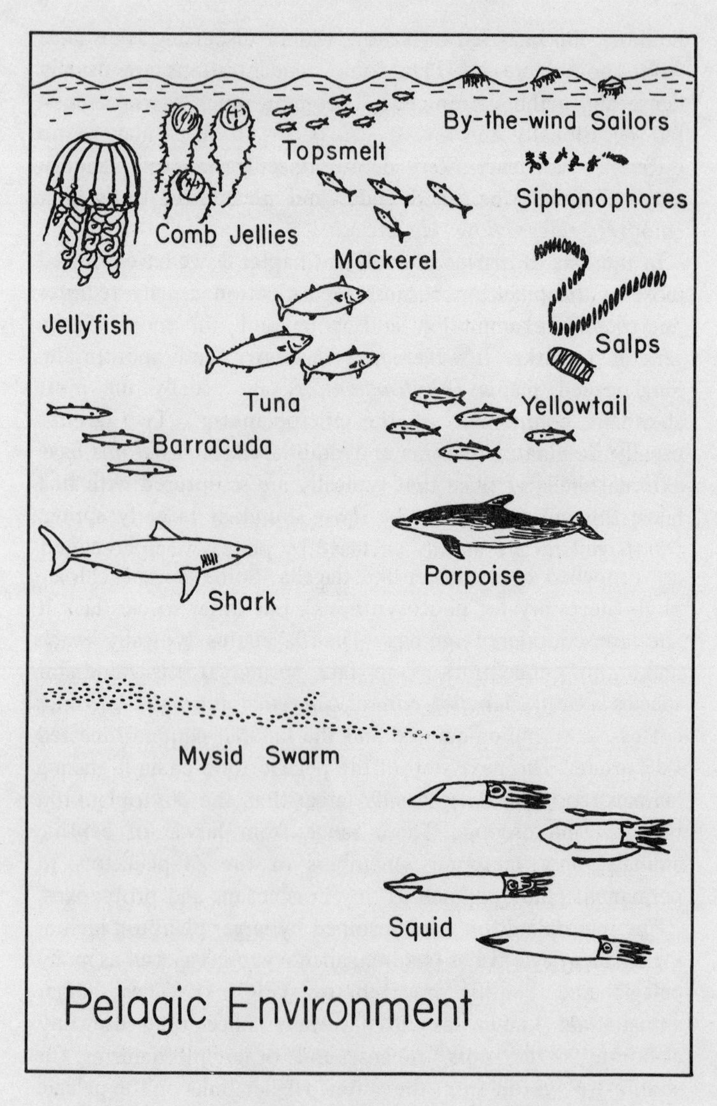

Fig. 13. Diagram of the pelagic habitat, with a few typical animals.

animals, the baleen whales (Gray Whale and others), are fairly low on the food chain. These whales are filter feeders, capturing plankton somewhat larger than the microscopic level.

Occasionally an uncommon pelagic animal will abruptly appear in enormous concentrations (jellyfish, By-the-wind Sailors, or even large nekton such as sharks or other fishes). The sea is suddenly swarming with a species that may have been completely absent for ten years. Such events can happen in several ways. Possibly conditions for growth of the large population were ideal in some part of the ocean, allowing the massive numbers to develop. An unusual current might then carry the population into the California region, and perhaps other mechanisms such as wind-induced currents might further concentrate the animals. For nektonic animals, an unusual abundance of the natural food might attract them to an area.

You will always encounter pelagic organisms on a dive, but if you wish to observe and study the larger planktonic animals it is better to do so several miles offshore, in the "blue water." Although this will require a boat, you will not be bothered by the turbidity that often impairs underwater visibility near shore. Usually the larger plankton such as ctenophores, medusae, tunicates, etc., are rather sparsely distributed. Good visibility considerably increases the chances of encounter. Although little may be visible topside from the boat, do not assume that the waters are barren. Many of the pelagic creatures are transparent or camouflaged so that one must be underwater to see them well.

SUBMARINE CANYONS, SEA CLIFFS, AND PINNACLES

For some reason, large vertical rock faces are nearly always rich in life and invariably worth a dive. Probably many factors contribute to species abundance and diversity here. Filter feeders on cliffs have access to water that has not been repeatedly strained by other benthic communities. Strong currents sweep these substrates, creating turbulence that enhances oxygen transport and removal of waste products.

SUBMARINE CANYONS

Nearly a dozen submarine canyons occur in California waters. Several are easily accessible without the use of a boat (La Jolla, Monterey, and Carmel canyons). Most have not been explored in any detail.

Walls of our submarine canyons vary from muddy gentle slopes to spectacular cliffs hundreds of feet high. The canyons themselves range from modest valleys to the mighty gorge that bisects Monterey Bay and compares in size with the Grand Canyon of Arizona. The canyons generally discharge into very deep basins; their upper edges are therefore usually nearer to deep water than other areas within the scuba zone. Deep-water animals frequently wander along canyon rims up to levels accessible to divers.

Like other areas of the continental shelves, the rocky parts of submarine canyons are usually more exciting than the slopes of sand and mud. Some of the steeper sediment slopes are almost totally barren because the substrate is "on the move"–gently in transit from the continental shelf to the deep sea. Some of the flatter sedimentary areas, however, may have fine concentrations of delicious clams or unusual creatures of interest to collectors or photographers. Hence the sand environment should not be ignored altogether.

A great deal of organic debris carried by currents along the continental shelves is intercepted by the submarine canyons and nourishes the resident fauna. Fresh and decomposing seaweeds accumulate in the axes and tributaries of canyons. This mechanism for concentrating food adds to the richness of life.

For successful canyon diving the arts of underwater orientation and navigation should be mastered, as well as the technique of taking topside bearings. The edges of canyons are usually so narrow that positions must be accurately determined–slight miss, and one can descend into dangerously deep water. The contours of some canyons are quite complex, and it is easy to lose one's way. If possible, examine a chart of the area while planning your dive, and become familiar with its contours. When you find an interesting spot, mark it with good bearings to facilitate returning another time.

SEA CLIFFS

Underwater cliffs range from isolated drop-offs a few yards long and a few feet high to formations several miles long and 30 or more feet (9 or more m) high. The more extensive cliffs often represent ancient shorelines formed when sea level is known to have been lower. Plant life disappears at much shallower depths on sea cliffs than on horizontal bottoms. Vertical walls are always shaded for at least a portion of the day, and this may affect survival ability of the flora on cliffs.

The base of a submarine cliff is usually buried in rubble which includes fragments of rock that have broken off the escarpment as well as remains of plants and animals that formerly inhabited the upper portions. There is also a tendency for drifting debris to concentrate near the base. The region of the rubble pile, therefore, is apt to be rich in life and an interesting place to collect. Don't neglect to turn over boulders and study the fauna beneath.

PINNACLES

Underwater pinnacles tend to be common off rugged coastlines dotted offshore with islets. They are rare where plains or gently rolling hills adjoin the shore. They are often remains of lava intrusions or similar hard rock. The softer surrounding material formerly enclosing the hard stone erodes first, leaving the isolated pinnacle. To be biologically interesting, pinnacles should be five or ten feet (1.5-3m) in both height and diameter. Many are much more spectacular–more than 100 feet (30.5m) high and several times this size in circumference. Some rise to fairly sharp peaks, others have flattened or truncated tops. Some are roughly circular in horizontal cross-section, others are elongate. Nearly all have an intricate topography and are interlaced with fissures, holes, and crevasses, displaying crags, overhangs, tunnels, etc. The same factors that encourage biological luxuriance along sea cliffs also operate to enrich pinnacles. In southern California and Baja California, some of the larger pinnacles support a lovely hydrocoral (*Allopora*) that grows in no other environment there, although it is quite common in northern California.

ENVIRONMENTS ALTERED BY MAN

Most of underwater California, as of this writing, is still in a fairly pristine condition. Certain species such as the sea otter and the gray whale are probably far below their former abundance, and undoubtedly populations of the more prized fishes and shellfishes are modified. The numbers of most plants and animals, however, and the basic structures of the communities, seem to be close to normal along most of the coastline. Nonetheless, exceptions exist.

Notable attempts have been made to improve our underwater seascapes. Innumerable conservation measures are promulgated by the Department of Fish and Game. The Department of Parks and Recreation is also deeply involved in conservation work and has recently begun to establish a system of underwater parks. Many artificial reefs to encourage fish and invertebrate populations have been funded by the Wildlife Conservation Board and constructed by the Department of Fish and Game.

On the other side of the ledger, mistakes have been made, largely through ignorance of the profound effects man can cause in the undersea world. State- and federal-sponsored research programs have been able to identify causes of trouble in certain cases and to apply corrective action. Nonetheless, human usage of our underwater is increasing very rapidly. It is by no means certain that man's ingenuity at conservation will be able to keep pace with his needs for ocean resources, to prevent irreparable losses.

One of the most widespread modifications in underwater habitat presently occurring is the conversion of bays and estuaries into harbors. Many bird species are affected immediately. There is considerable ignorance about long-term effects on subtidal invertebrates, fishes, and plants. Certainly areas such as Mission Bay and lower Newport Bay are relatively sterile at the present time, compared either to their condition before extensive development or to fairly untouched regions such as Bodega Bay. Harbor development in other locations has had even more serious consequences due to industrial pollution. Usage for heavy shipping and waste disposal has adversely influenced biota of San Francisco, Los Angeles, and San Diego bays. Corrective action has been taken in many areas; undoubtedly conditions in some of these areas will improve.

Disposal of collected sewage and industrial wastes into the open sea has been practiced in California since before the turn of the century. Effects appear to have been negligible before the great population expansion of World War II. Since then, however, certain kelp areas in southern California have declined drastically, and the urchin-dominated environment has become very widespread near Los Angeles and San Diego.

Considerable improvement in the design of sewer outfalls has taken place more recently, and most wastes are now discharged below the scuba zone. Dr. David Leighton at the University of California's Institute of Marine Resources discovered that urchins can be selectively controlled by chemical or physical means. These techniques are currently being used by the Kelco Company of San Diego to restore kelp beds which they harvest.

Concern has been voiced that heated water discharged to the sea by electricity-generating plants will affect marine communities adversely. At present only a few such discharges exist. Discharge temperatures are 8° to 16°F (5°-10°C) above ambient and cause changes only within quite a small area. With the advent of nuclear power plants, much bigger installations can be expected. As needs grow, discharges might be spaced ever more closely. Nonetheless, "thermal pollution" in the sea should only be a minor problem for many years to come. If major problems ever do arise, the power industry can adopt some of the extremely efficient dispersal systems now in use by the sewage-disposal plants.

Effects of gross pollution are easily perceived. The bottom may be completely devoid of life or there may be only a few species in an environment that would normally support dozens. The numbers of organisms can be deceiving. Sometimes a grossly polluted situation will support thousands of animals representing only one or two species that thrive in the modified environment.

The early or milder stages of pollution can be difficult to recognize, inasmuch as changes tend to mimic alterations that occur naturally. The natural modification, however, typically reverts back to the initial state, whereas effects due to pollution persist and may be aggravated as an area receives increasing volumes of waste. Minor changes in frequency and abundance of species are typical manifestations of mild pollution. Detection requires considerable familiarity with the normal condition.

COMMUNITY TYPES

SUBMARINE FORESTS

Submarine forests are scattered along the entire coast of California and northern Baja California. The large brown kelps are the "trees" in these forests (Fig. 14). Five species are involved: Giant Kelp (*Macrocystis*), Bull Kelp (*Nereocystis*), Feather-boa Kelp (*Egregia*), Elk Kelp (*Pelagophycus*), and a type of sargassum weed (*Cystoseira*). All depend upon gas-filled floats (*pneumatocysts*) to buoy up the *stipes* (stems) and *blades* (leaves). Bull Kelp is the most important plant north of San Francisco. Between San Francisco and Point Conception, the forests may be pure stands of Bull Kelp, of Giant Kelp, or of sargassum weed, or mixtures of the three. Sargassum weed and Giant Kelp often predominate along the inshore borders, with Bull Kelp more abundant offshore. South of Point Conception, Feather-boa Kelp dominates the inshore borders and Giant Kelp occurs at intermediate depths; south of Point La Jolla, Elk Kelp usually lines offshore borders.

Fig. 14. Diagram of a grove of giant kelp, *Macrocystis*, representative of a submarine forest habitat.

The tree-like kelps are well adapted to survive in this environment, where every passing wave generates hurricane-like forces against all exposed surfaces. If these plants had large ponderous trunks like terrestrial trees, storm swell would rip them out just as high winds topple large unbending structures on land. The flexible grasses survive hurricanes, however; and submarine trees behave similarly. They yield and bend in wave surge, even though the stipes are very tough and difficult to break by stretching. Another important adaptation displayed by several kelps is the ability to translocate photosynthesized material from the surface blades down the stipes, to support growth of organs in the dimly illuminated bottom environment. Light measurements by Dr. Michael Neushul of the University of California proved that without translocation, lower portions of kelp plants would starve in the thicker kelp beds. The very existence of a forest type of habitat requires some kind of translocation mechanism to cope with self-shading. Dr. Bruce Parker and others demonstrated the existence of translocation by following the movements of radioactive compounds they "fed" to kelp plants.

Like their terrestrial counterparts, submarine forests are a haven for many animals. The kelp fronds form intricate crevice environments where juvenile and small animals can hide from predators. Many animal species graze on the kelps; the forests produce a great deal of plant material and are thus able to support huge populations of grazers. The enormous surface created by all the blades in the forest provides living space for the multitude of tiny encrusting organisms that otherwise have only rock surfaces available to colonize. According to the late Dr. K. A. Clendenning of the University of California, an average kelp bed furnishes fourteen times the encrusting surface provided by the bottom alone.

Anyone who has had to scrape a boat bottom or dock knows that encrusting organisms can quickly build up to large unwieldy masses. Such growths on plant surfaces could soon weigh enough to counteract pneumatocyst bouyancy, sinking fronds and destroying the forest. Kelps have adapted to cope with this problem by rapidly turning over their tissues. Fronds of Giant Kelp, for example, become senile in six months; the individual blades may last only a month or two and then deteriorate. The large heavy encrusters such as barnacles and mussels simply do not have sufficient time to develop. Some encrusters can cope with kelp's fleeting existence, however, and do so by having even shorter developmental times. The membraniporan bryozoans, for example, can completely overgrow a kelp blade in three weeks. These animals have adapted by reproducing copiously and forming colonies. Each animal is microscopic, so it does not require a long growing period before reaching adulthood. Although a colony covers considerable area, not much animal tissue is involved because the encrustation is quite thin; this makes the encrustation semitransparent, causing little light absorption, while colony weight remains light, rarely sinking the frond. Mature bryozoan colonies in turn serve as food for other organisms. Because bryozoans eat microplankton, the colonies link kelp-community food chains to phytoplankton production. Existence of the bryozoan-kelp association can thus substantially enrich life in kelp beds and is made possible by specialized adaptations characterizing the colonies.

Giant Kelp and Bull Kelp can grow in such dense concentrations that blades near the surface (the *canopy*) block out more than 99 percent of the sunlight. Most plants cannot grow in the dim light beneath a kelp canopy. They are replaced largely by encrusting animals—the sponges, hydroids, bryozoans, and tunicates. Shade-loving motile animals (lobsters, many fishes) are attracted by the darkened forest environment. The floor of a kelp bed thus tends to resemble the deeper rocky bottoms of the scuba zone.

Submarine forests differ from terrestrial forests in their persistence. Forests of the land last for decades, sometimes centuries. Forests of the sea, however, may last only a year or two. Each winter the stands of Bull Kelp are almost entirely ripped out by storms. Beds of Giant Kelp usually last several years, but they too come and go. Armies of grazing animals, particularly sea urchins, can remove all vegetation from many acres of the sea floor; this barren condition will persist until the grazers either move elsewhere or die of starvation. Eventually plants become re-established, and a new submarine forest develops. The complete cycle may require 5 to 10 years or even longer.

Submarine forests are usually excellent diving areas. The jungle-like environment with its wealth of animal life provides recreation alike for the hunter, the photographer, and the curiosity-seeker. Like a jungle, however, submarine forests have their hazards. Kelp-bed animals are not dangerous, but the plants can cause difficulties. Underwater, the stipes become entangled around equipment; at the surface, dense canopies can impede swimming. It is best to avoid the thick bundles of stipes in midwater and seek passage through the relatively open spaces that occur near the bottom. If it is necessary to swim through a thick canopy, do so slowly, keeping as horizontal as possible. Press the kelp fronds under as they approach, so your chest slides over them. Do not thrash aimlessly if you become entangled. Find the critical stipes and break them one at a time by bending (not by stretching, as they are slippery and elastic).

Kelp in southern California is harvested and used as fertilizer and food additives, and processed for chemicals. Many years ago entire plants were ripped out by the harvesting method, but modern techniques cut only the tops off. The cutting produces little, if any, effect on the kelp plant. Canopy removal also occurs during severe storms, so that present harvesting methods tend to mimic a natural process.

SHRUB COVER

Some seaweeds shorter than the Giant Kelps are nevertheless sufficiently tall to form a mass of plant material anywhere from 1 to 6 feet (2.5cm-2m) above the bottom. The tops of these plants never reach the surface except in shallow water at low tide. These midwater canopies have several features in common with the habitat formed by surface canopies of submarine forests. The shrub cover is used for shelter, food, and substrate by most of the animals that similarly use the submarine forests. Shrub cover also shades the underlying region, sometimes as effectively as canopies of the submarine forests.

Shrub cover is formed principally by brown algae (usually the palm kelps). Red algae will occasionally become sufficiently tall and massive to contribute to this community type. Eel Grass in bays can be considered as shrub cover, and the Surf Grasses in the open sea act partially as shrub cover, though narrowness of the blades restricts encrustations and grazing. Shrub cover usually begins about a foot below mean sea level and extends out to the edges of submarine forests, if any. Sometimes a moderate shrub cover occurs within submarine forests when surface canopies are light.

The "trunks" of palm kelps represent an adaptation to enable these plants to function as "shrubs": they can shade out the low-statured plants and sometimes even prevent formation of submarine forests by hindering development of juvenile Giant and Bull Kelp. The palm kelps do not have pneumatocysts, and the thickened trunk-like stipe is used to hold blades aloft. The stipe has some flexibility to survive wave surge, and elasticity is well developed, so that the structure quickly straightens after a wave passes. Magnitude of wave surge decreases with depth, and palm kelps rarely exceed heights of 2 to 4 feet (61cm-1.5m) so the "trunk" method of supporting upper foliage is satisfactory for such plants in deep water. A few stout shrub-like species can even survive in the intertidal region.

Learn to interpret the shrub cover. It quickly reveals whether an area is typically a warm-water or a cold-water region. In central and southern California and Baja California, the Southern Palm Kelp (*Eisenia*) predominates in warm-water shrub cover. Moderately cold water will encourage fairly pure stands of Northern Palm Kelp (*Pterygophora*). In the coldest regimes, several blade-kelp species (*Laminaria setchellii* and others) mingle with Northern Palm Kelp.

In many places large sessile invertebrates may become so numerous that they form an effective shrub cover; gorgonian coral stands are the commonest example. Such a shrub cover of a mixture of invertebrates sometimes stretches more than a foot down from overhangs.

TURF

Terrestrial turf is a mat 1 or 2 inches (2.5 or 5.1cm) thick covering the ground, composed of an interwoven mixture of plant stems, roots, and soil. In the sea, most rocky bottoms support a similar turf. Unlike the land turfs, skeletal structures of invertebrates often contribute significantly to the cohesiveness of submarine turfs. Sometimes surfaces will be covered exclusively with invertebrate turf (e.g., a wharf piling community). We will identify major types of turf as plant, mixed, or invertebrate, according to primary compositions. In southern California and Baja California, turf is usually low and distinct from shrub cover. In central and northern California, however, plant turfs can become so dense and deep that they grade into the shrub cover and may even crowd out the palm kelps (e.g., stands of surf grass or coralline algae). Because animal turfs are not limited directly by light, they usually proliferate in dim locations–beneath submarine forests and overhangs, in deep water, turbid places, caves, etc. Thus walls of submarine canyons are often covered with encrusting sponges, bryozoans, and tunicates. Short plants dominate in well illuminated environments, and may inhibit or modify colonization by invertebrates.

Turf-dwelling organisms continually face being crowded out or covered up by other members of this community type. The problem is solved through various adaptations and mechanisms by different species. Some organisms such as sponges presumably eat any tiny larvae that approach their surfaces, sweeping the microscopic creatures into a digestive cavity by ciliary currents. Bryozoans often have beak-like avicularia to pluck off any object that settles on the colony. Some plants are suspected of exuding antibiotics to inhibit would-be colonizers. Some compound tunicates have soft gelatinous surfaces that presumably are poor substrate for larvae preferring something solid. Other creatures tolerate some colonization for camouflage. Certain spider crabs actually "plant" bits of sponges and algae atop the carapace.

Generally, more species dwell in the turf than in any other community type. It is not uncommon, however, to find large patches of turf formed exclusively of one or two species (e.g., mussel colonies, clusters of tube mollusks, beds of short red algae). Sometimes what appears to be a single-species community proves on close examination to consist of many organisms depending upon the conspicuous dominant in some way or other. Thus dissection of layers of sponges may reveal many tiny commensals within the chambers.

Invertebrate turfs occur occasionally on sandy bottoms. They usually contain one or two dominant species that bind the mass into a unit; tube-building worms such as *Owenia, Diopatra*, or *Chaetopterus* are often the binding ingredient. Even fairly loose-lying aggregates of sand dollars, sea pansies, or bivalves can attract other organisms in sufficient quantities to form a turf.

THE URCHIN-DOMINATED ENVIRONMENT (Plate 7c)

If you dive in rocky areas several times, the chances are good that you will soon encounter the urchin-dominated environment. The sea urchin—a spiny, roundish animal that resembles a pincushion—at times becomes so numerous that its normal food supply of seaweeds is entirely consumed, exposing the underlying rocks. Urchins appear to be able to survive for very long periods under these barren conditions. Many other animals, however, also depend on seaweeds. They do not have the lasting power of the urchins, and these other grazing species either starve to death or migrate to undecimated areas. Urchins become carnivorous when they are hungry. After they have destroyed most of the plant life, they consume the more delicate sessile filter-feeders. Eventually all that is left besides urchins are species with very hard shells, including certain coralline algae.

After urchins become dominant, events may follow one of several courses. Sufficient drifting debris may be swept into the area to prevent total starvation, and the urchin hordes persist. The urchins may also move to greener pastures, or eventually they too may starve to death. If, by one means or another, urchins disappear, seaweeds start to return in a matter of weeks and attain their former luxuriance in several months. The associated animal populations, however, may require several years for complete restoration. The cycle may repeat itself if sufficient urchins migrate back, or if currents bring large numbers of urchin larvae into the area just as they are settling from the plankton.

Ravages by urchin swarms are probably only a fairly recent phenomenon. An important urchin predator, the sea otter, was exterminated from nearly the entire California and Baja California coast by fur traders in the nineteenth century. The otter herds have returned, however, in parts of central California. Otters have moved in where urchin-dominated environments once existed in these regions, and the only urchins that can be found now are far back in crevices where otters cannot reach. Rich seaweed stands have developed in these areas, so it appears that the urchin plagues have been successfully controlled. The otter herds are growing slowly; perhaps some day they will again be able to restore the balance of nature along our entire coast.

Chapter 5

SOME EXCELLENT DIVING SPOTS

Eight areas will be described below. This, however, does not imply that other places are inferior. A book ten times the size of this volume would be needed to discuss all the wonderful diving locations in California waters. Our eight choices were based on the following general criteria:

All the selections are spots that are teeming with plants and animals. Several have been popular diving places for years; hence edible species may be drastically reduced. Nonetheless, the spots all support abundant nonedible species and have not yet been picked clean. All are places with a variety of habitats and communities, and in most cases one or more of the habitats is outstanding in some respect. The locations generally have easy access to the water and usually contain one or two spots where safe entries are possible in all but the worst weather. Many are near harbors, with boat rentals and launching facilities available. The selections were intentionally spread throughout the state. All represent open-sea habitats. Bays and estuaries were not included, because they are quite specialized situations and to do them justice several should be discussed, scattered along the entire coast. Space limitations prevented inclusion of all place names on the accompanying charts. The names are often on road maps or you can inquire as to locations at a local diving store.

1. PUNTA BANDA (Fig. 15)

Location: Punta (Spanish for "Point") Banda forms the southern edge of a large bay, Bahía Todos Santos, in Baja California. After crossing the U.S.-Mexico border, take Highway 1 to Ensenada and continue south about 10 miles (16km)

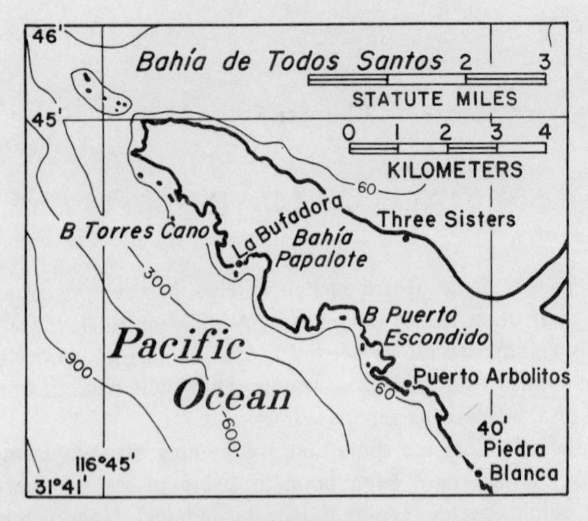

Fig. 15. Chart of the Punta Banda peninsula near Ensenada (see Fig. 8), showing locations described in the text.

to a village called Maneadero. If you wish to proceed southward beyond Maneadero, it is necessary to stop for immigration inspection at a station on the southern outskirts of town and submit a visa (visas are free and can be obtained at offices of the Mexican Consul or the Mexican Tourist Bureau in Los Angeles, San Diego, or Ensenada); proof of United States citizenship is required. No visas are needed, however, to visit Punta Banda. The highway forks in Maneadero; take the right fork, a paved road to Punta Banda. Two to three hours of driving from the border are required to reach the area. A Mexican sportfishing license is necessary to dive–it can be obtained in sportfishing stores in Ensenada, and costs 50 cents for three days.

Services: Diving stores are located in Ensenada (with air refills) and at Papalote Bay (La Bufadora) on Punta Banda. There are restaurants and grocery stores on Punta Banda at Papalote Bay and Camp La Jolla. Campsites are scattered over Punta Banda, trailer parks available at Camp La Jolla. There are boat-launching ramps at Camp La Jolla and a crude ramp at Papalote Bay. Boats can be rented at Camp La Jolla, Three Sisters, and Papalote Bay. Motels available in Ensenada.

Access: Three Sisters and Camp La Jolla (all weather) from the north side; Papalote Bay and Abalone camp (nearly all weather) from the south side.

Punta Banda is largely rugged, undeveloped wilderness. Access is limited by terrain rather than by private ownership. Many access spots are available via hiking trails, but are difficult and strenuous. Land orientation is such that with bad weather from the south, the north side is calm and good diving; and vice versa for bad weather from the north.

GENERAL DESCRIPTION

Punta Banda is about 5 miles (8km) long and 1 mile (2km) wide. The north side borders a sandy bay, but most of the beaches and shallow subtidal are rocky. The south side is a maze of rock-bound inlets and small islands with occasional sand patches, dropping steeply to a sandy shelf about 130 feet (39.5m) below the surface. Wind regimes cause intense upwelling on the south side, and the water is always cold and generally clear. A paved road passes along the north side, crosses the headland at about the midpoint, and ends on the south side at a large inlet, Papalote Bay.

Best diving is found at the western end of Punta Banda, where a long chain of islets marks a rich underwater ridge. This is accessible only by boat, however, but excellent areas exist in Papalote Bay and elsewhere, off beaches that can be reached by auto.

POINTS OF INTEREST

SANDY ENVIRONMENT

Most of the north side of Punta Banda, at distances of 100 to 200 feet (30.5-61m) from shore, is fine sandy bottom. Coarse sand occurs in the central parts of Papalote Bay and in the centers of major inlets on the south side.

ROCKY SHELVES

Narrow and gently sloping rocky environments border much of the north side. A rocky shelf about 2000 feet (610m) broad extends out from the south side.

PINNACLES AND CLIFFS

The south side is outstanding in its complement of pinnacles and underwater cliffs. Pinnacles are scattered from the tip of Punta Banda eastward to the center of Papalote Bay. Almost any islet well out from shore is the top of a pinnacle. Papalote Bay is one of the few places where good pinnacles lie within swimming distance from shore.

There are several sea cliffs in Papalote Bay, and probably many more await discovery along the south side. The outer western edge of Papalote Bay is a luxurious sea cliff. The outermost island in Papalote Bay marks the top of a small but interesting sea cliff.

SUBMARINE FORESTS

Fringing kelp beds line much of the rocky shore where some protection from storm waves occurs. Larger beds exist in Papalote Bay and in Todos Santos Bay near the base of Punta Banda. At times a moderate kelp bed develops on the rocky shelf just south of the tip of Punta Banda. Kelp often grows on the tops of submerged pinnacles and facilitates finding them. A small roundish kelp patch well out from shore is apt to be a pinnacle.

The Punta Banda forests are especially fine for photography. Only a few islands have water as clear as Punta Banda, and it is often possible to photograph entire plants with their associated organisms. The record depth of 132 feet (40m) for Giant Kelp was recorded at Papalote Bay.

The cold upwelled water on the south side furnishes a wealth of nutrients for plants, and very luxurious growths are common. Coldness also encourages the more northern species, and submarine seascapes here are reminiscent of the bottom around Monterey or farther north. The cliffs and pinnacles are usually covered with colorful mats of sponges, mollusks, and tunicates that are absent or poorly developed in southern California. Plush carpets of wine-red algae clothe the tops and upper sides of many pinnacles. Hydrocorals encrust the midsections of some of the outer pinnacles, and sheets of glowing fluorescent anemones cover many of the vertical cliffs. On the flatter shelves in deep water, some of the magnificent foliose red seaweeds are well developed. Often the blades are 3 or 4 feet (1 or 1.5m) long, so thin that they are transparent. They stream out gracefully in the gentle surge caressing the bottom. In places the bottom is hidden by clusters of multicolored brittle stars. From a distance, the aggregates resemble brilliant Persian rugs carelessly draped over the boulders. The excellent visibilities allow observation of shy fishes that always keep a good distance away from humans.

URCHIN-DOMINATED COMMUNITIES

The bases of many pinnacles and some of the rocky shelves at Punta Banda are dominated by urchin populations. A few other organisms occur adventitiously in these areas (starfish, anemones, gastropods), but they are generally sparse. Only one type of plant survives the voracious grazing activities of the urchins—the encrusting coralline algae. The stems and upper parts are usually grazed away but the pink-to-purple flat bases remain. The excellent water clarity allows the corallines to extend very deeply. The black or blue urchins stand out vividly against the gray-pink skin of corallines that covers all rock surfaces in this environment.

HOT SPRINGS AND THE BLOWHOLE

Underwater hot springs are situated in Papalote Bay as well as in the estuary on the north side. Flow from the Papalote Bay spring issues over roughly half an acre at depths of 80 to

100 feet (24.5-30.5m). Small streams of bubbles also come up through the sand. A shimmering effect can be seen near the bottom, similar to shimmering seen above a hot pavement in summer. The heat flow is not sufficient to affect water temperatures significantly, but the bottom itself is appreciably warmer. Temperatures are sufficiently hot to cause pain a few inches beneath the sand.

La Bufadora, the blowhole, is a geyserlike spout of seawater energized by the waves. Although not an underwater phenomenon, it is well worth taking a few minutes to see. La Bufadora is at the end of the paved road in Papalote Bay. No one knows how the energy is channeled to hurl great quantities of water 100 feet (30.5m) into the sky. The risks are too great for divers to investigate. It is safe, however, to explore a delightful little underwater valley that runs seaward from the entrance to La Bufadora, out to the sea cliff.

2. LA JOLLA (Fig. 16)

Location: The resort town of La Jolla lies at the southern end of the Gulf of Catalina. It is about 100 miles (161km) south of Los Angeles and 15 miles (24km) north of San Diego. Centuries ago a stream probably flowed from the hills onto the flatlands now called La Jolla Shores. At the turn of the century the site of the flatlands was a marsh-lagoon environment. Landfills have reduced the former lagoon to a small pond behind the Beach and Tennis Club, and residences now overlie the old marshes. Indian villages were located at La Jolla as far back as 7000 years ago. The area is easily reached by turnoffs from Interstate Highway 5, either from the north or south.

Services: Hotels, motels, and restaurants are plentiful in La Jolla and nearby (less expensive) sections of San Diego such as Pacific Beach and Mission Beach. Diving stores with compressed air and equipment rentals are located in La Jolla and Pacific Beach. Boat rentals and launching facilities (ramps and

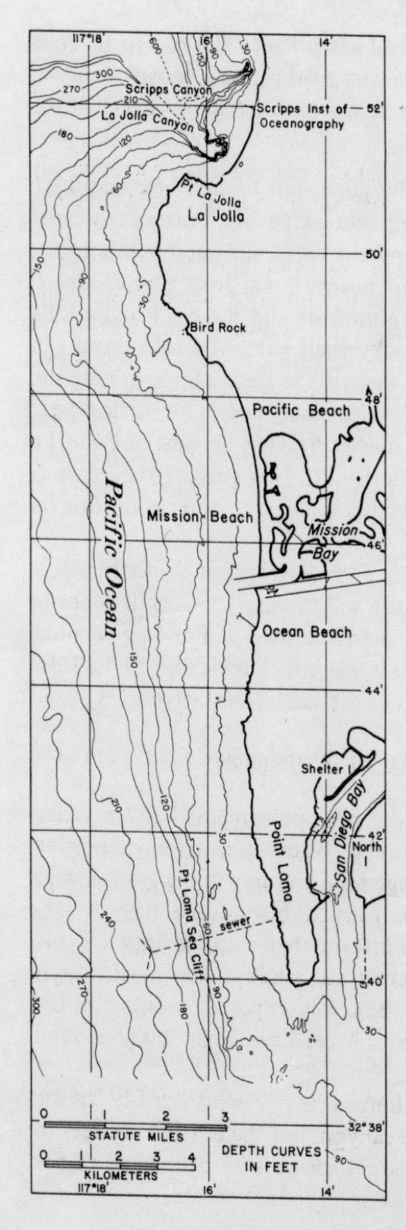

Fig. 16. Chart of the coast from La Jolla Bay to Point Loma, showing bottom contours and the La Jolla submarine canyon.

hoists) are available in Mission Bay Park. The run to La Jolla from Mission Bay is about eight miles (13km), mostly exposed sea. Skiffs can be launched across La Jolla Shores Beach at the end of Avenida de la Playa.

Access: Access is generally good, but limited here and there by impassable cliffs or by private ownership. Fortunately most of the seaside homes occupy the cliffs, and good public access is available from the open beaches. La Jolla Shores Beach allows access to sandy environments and the head of La Jolla Submarine Canyon. The rocky shelf below La Jolla Caves can be reached by trails down the cliffs at Devil's Slide or near the Cave Store. The Cove provides good entry for diving near Point La Jolla. Numerous pocket beaches lie west of Point La Jolla and give access to the water. This part of the coast is exposed to the open sea, and it is wisest to enter and leave via sandy areas.

Point La Jolla protects La Jolla Bay against southern swell. During bad weather from the south, enter at Casa de Manaña breakwater, the Cove, or points eastward. For bad weather from the north, the only safe place for diving may be La Jolla Shores Beach near the head of the submarine canyon.

GENERAL DESCRIPTION

La Jolla Bay is one of the few semi-enclosed inlets facing north in southern California. Water circulation tends to produce warmer-than-average temperatures in the Bay. A large and fascinating submarine canyon bisects the Bay. It was apparently formed by the stream that emptied into the old lagoon. Several thousand years ago sea level was much lower, and most of the present Bay was exposed except for the canyon. Remains of Indian villages have been found around the edges of the canyon.

East of the canyon, the bottom is fine sand and silt. To the west, fine sand borders the canyon, but the bottom changes to rock as shore is approached. Rock persists over most of the broad shelf that lies off western La Jolla.

POINTS OF INTEREST

SANDY ENVIRONMENT

Excellent examples of sandy-bottom communities occur along the east side of La Jolla Bay. Just north of Scripps pier, a dense bed of sand dollars (*Dendraster excentricus*) lies at depths from 25 to 40 feet (7.5-12m). Many other animals associate with the thick clusters of sand dollars. These modified urchins also appear elsewhere in the Bay at similar depths, but are usually not concentrated. Meadows of Eel Grass (*Zostera marina*) may be found out on the flats at 30 or 40 feet (9-12m), near the borders of the canyon. Presumably they are all that remain of larger Eel Grass stands that must have existed when the marsh and lagoon were present.

ROCKY BOTTOM

Most of the rocky shelf off La Jolla is pavement rock with isolated boulder patches and low-lying ledges. In the vicinity of Point La Jolla, however, bottom irregularities become prominent, and the terrain is quite interesting. There are large boulder piles, a few cliffs and pinnacles, and at least two underwater caves. The east side of Point La Jolla is influenced by the slightly warm temperature regime of the Bay, and cold-water species are rare or absent. A colorful appearance results. Swirling Surf Grass creates a green wavy carpet just below the intertidal. Slightly lower, the shrub cover of brown algae becomes prominent. These plants are light brown to yellow, not the somber browns common north of Point Conception. Dense tufts of pink coralline algae and some of the brighter reds form a turf beneath the shrub cover. A host of dazzling fishes hover above and around these pastures—Garibaldi, Kelp Bass, Sheephead, Rock Wrasse—adding their lovely colors to enhance the scenery.

Cold-water forms appear farther out on the shelf. For the most part they colonize floors of the kelp beds or deep-water cliffs. Gorgonian corals seem to proliferate in this area. At times, intensely cold water upwells from the nearby canyon and is swept across deeper portions of La Jolla shelf. This nutrient-rich water undoubtedly encourages richness of both flora and fauna.

SUBMARINE FORESTS

The rocky shelf supports superb submarine forests of Giant Kelp and Elk Kelp. Both these species are sensitive to warm water, so at times kelp inside La Jolla Bay is straggly or disappears altogether. Populations of the huge black sea bass and grouper formerly dwelt in the warm fringes of the La Jolla kelp bed. Nowadays only a few wary individuals have successfully eluded the throngs of spearfishers hunting in this famous area. When conditions favor kelp, the giant plants colonize the Bay almost as far as the Caves. A series of warm months, however, can cause recession southwestward along the shelf for 2 or 3 miles (3 or 5km). As of this writing, the kelp bed is recovering from the worst exposure to warm water on record. Water temperatures remained high for two years (1957 to 1959) and kelp recolonization has taken more than a decade. Restoration was also hindered by plagues of urchins. A program of urchin control was conducted by the local kelp harvesting industry (the Kelco Company of San Diego), assisted by several universities and diving clubs. Thousands of urchins were removed or destroyed, and the kelp has gradually returned.

LA JOLLA SUBMARINE CANYON

Most of the gentle upper slopes of the canyon are covered with sand that is slowly sliding into the steeper parts. The river of sand eventually flows down the axis out to the San Diego Trough. At times storm waves cause instability in large masses of sediment hanging near an edge, and underwater avalanches occur. These abrasive movements keep the canyon open and erode it deeper. The movement also creates an unstable environment and prevents the usual sand-dwelling organisms from becoming established. Only motile creatures such as flatfish and white urchins are common on the creeping slopes.

The richest places are rock outcrops and cliffs free from the smothering sand flow. Invertebrate turf is often richly developed here. It is certainly one of the outstanding examples of this community type on the entire coast. The turf attains thicknesses of 6 to 12 inches (15.2-30.5cm) along the upper rims of some of the great cliffs. Massive fish schools invariably associate with these dense living carpets, undoubtedly depending on them for food and shelter.

The canyon divides into two branches in its upper regions. The northeastern branch, Scripps Canyon, is a steep narrow gorge. The western trench, the La Jolla Branch, is much broader and more accessible, but has fewer rocky areas within the scuba zone. Even so, one or two of the cliffs here are unmatched by anything elsewhere in California waters.

Three fine diving spots in Scripps Canyon are known as The Junction, North Branch, and the Rock Pile. The Junction is a maze of hanging valleys and tributary small canyons, all converging on the main gorge at about the same place. One tributary is so precipitous that a diver at 90 feet (27m) can touch opposite walls of the canyon with outstretched hands, while the bottom lies another 135 feet (41m) beneath his fins. North Branch has the most shallow rock outcrops in the canyon, at depths of 50 to 100 feet (15-30.5m) and is one of the few places supporting seaweed growth. The walls of the canyon proper are vertical or undercut here and there to form big ledges and cavelike indentations. The Rock Pile lies on the west side, opposite The Junction. Massive slabs of rock have broken away from a thick ledge at the rim of the canyon. They have "stranded" while sliding down the sandy slopes, and other boulders have accumulated behind. The rose-pink gorgonian coral, *Lophogorgia*, flourishes on these slabs, and the area is also known as the coral forest. The Rock Pile is also outstanding for its nudibranch populations. Most of these sea slugs are closely associated with the coral, and probably use the polyps for food.

The portions of the La Jolla Branch nearest shore display small cliffs cut in soft mudstone at depths of 50 to 60 feet

(15-18m). Hosts of clams take advantage of the soft substrate, and the cliffs are riddled with their burrows. Many are edible varieties. Hundreds of primitive human artifacts have been recovered by divers from the southwestern edge of the La Jolla Branch. Radiocarbon dates place the age of these Indian campsites as 4000 to 7000 years ago.

Moving seaward, cliffs of the La Jolla Branch become larger, steeper, and formed of harder rock. The Cod Hole, opposite Point La Jolla, lies at about the maximum depth range that can be safely worked by divers using compressed air (130 to 200 feet [39.5-61m], it is not a place for novices). The uppermost rock outcrops begin at depths of 130 to 150 feet (39.5-46m). They mark the top of a 500-foot (153m) cliff. Species characteristic of great depths sometimes stray up into the scuba zone here. At times strong upwelling currents flow straight up the cliff and spill out of the canyon like a gigantic waterfall in reverse. The environment becomes intensely cold and perhaps encourages deep-water forms to migrate above their usual ranges. Like the Rock Pile, gorgonian corals dominate the landscape in the Cod Hole. The great cliff, however, supports several gorgonian species instead of just one. Cold-water nudibranchs are also common.

MISCELLANEOUS

The seaward edge of the rocky shelf off Point La Jolla drops abruptly 20 to 30 feet (6-9m) in several places down to sandy bottom. The area probably represents a fragmented sea cliff. The pinnacles and clifflets support a lush invertebrate turf and attract large fish schools. One area, Quast's Hole, displays a small cave at the top of a pinnacle. The cave has a hole or skylight in the roof.

Save time to see the museum at the Scripps Institute of Oceanography. Admission is free, though there is a nominal parking charge. Displays illustrate many fascinating aspects of the science of the sea, and an excellent aquarium exhibits many of the local fish and invertebrate species. The Scripps campus is beautifully designed and landscaped and worth a stroll. Regrettably, the laboratories and supporting facilities are, of necessity, closed to the public.

3. WHISTLER REEF (Fig. 17)

Location: Whistler Reef is about a mile and a half (2km) easterly from Newport Harbor and 3 miles (5km) westerly of Laguna Beach. The nearshore region is part of the Irvine Coast Marine Life Refuge. There are many fine reefs in the Laguna-Newport area, and Whistler Reef is undoubtedly one of the richest. The center of this shoal lies off a minor bend in the coastline known as Pelican Point. Pelican Point is reached directly by automobile via California State Highway 1. The

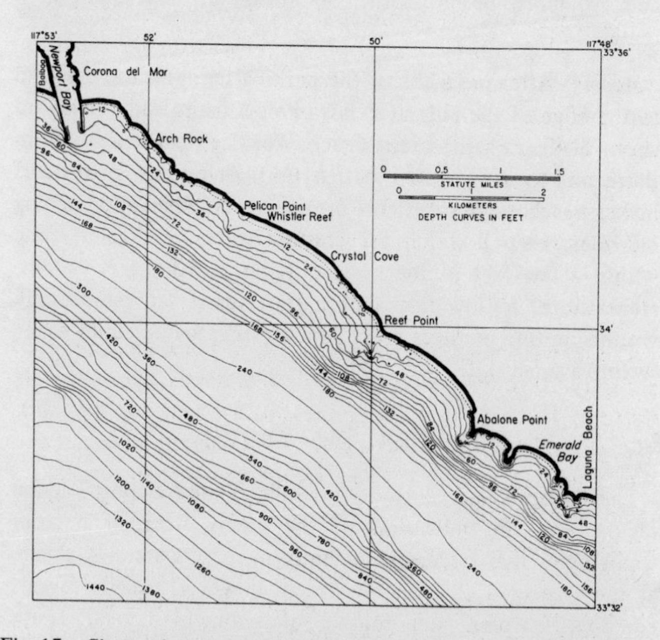

Fig. 17. Chart of the coast from Newport Bay entrance to Laguna Beach. Whistler Reef extends for about 1000 feet southwesterly, commencing at Pelican Point, from the shore out to 50-foot depths.

turnoff is about half a mile southeast of Corona Del Mar; horse corrals and stables mark the area. Parking and use fees are $1.00 per car.

The Reef is a bit hard to locate when approaching by boat, because Pelican Point is fairly inconspicuous from the sea. At the east edge of Corona Del Mar the beach changes abruptly from rock to sand. A half mile farther, at Pelican Point, it reverts to rock once more. The changes are easily seen by running from 500 to 1000 feet (153-305m) from shore.

Services: Hotels, motels, and restaurants are abundant in the nearby resort areas of Laguna Beach and Newport Beach-Costa Mesa. Boat launching ramps as well as skiff rentals are present in Newport Beach. Diving stores are located in Laguna Beach and Newport Beach, with compressed air and equipment rentals.

Access: After payment of the parking fee, you can drive up to the edge of the cliff at Pelican Point. Trails lead downward about 50 feet (15m) to the beach. Whistler Reef extends from shore out to about 2000 feet (610m). At times a strong surf makes beach entries difficult or hazardous. When approaching by boat, keep a watch for cresting waves. Pinnacles come within a few feet of the surface as far as 1000 feet (305m) from shore. At low tide a deep-draft vessel can temporarily ground on one of these crags, and cresting waves could easily swamp a small skiff.

GENERAL DESCRIPTION

The flatlands and gentle rolling hills of the Newport-Laguna region give no indication of the highly irregular bottom topography that occurs abundantly in these waters. A glance at the face of a cliff on the beach, however, usually reveals complex warping and folding of sediment layers. These intricate patterns account for most of the unevenness found underwater. The sea has dissolved or eroded away the softer sedimentary layers, leaving weird-shaped hard layers projecting above the sea floor as cliffs, pinnacles, crags, overhangs, and other grotesque formations. A great variety of micro-environments are thus created.

Fauna and flora of the Laguna-Newport area have a slight tropical affinity, although the water is usually cold and orientation of the shoreline favors upwelling. The region may occasionally be influenced by the northward-flowing Davidson Current, bringing in subtropical species such as grouper and the long-spined urchin, *Centrostephanus*. Plant life appears to be modified. Seaweed turf and shrub cover are reduced below 40 feet (12m) and often totally replaced by a rich invertebrate turf. The large foliose red algae are generally absent. The only abundant plant community present is the submarine forest. Relative scarcity of plants below 40 feet (12m) may be due to widespread populations of urchins, or it may result from turbid water flowing from nearby Newport Bay.

POINTS OF INTEREST

SANDY ENVIRONMENT

Fine-grained sand occurs near shore. Fragments of invertebrate skeletons and coralline algae become common below 20 feet (6m) and the sediments coarsen. This coarseness persists for a few hundred feet beyond the outer boundary of the reefs. Finally at depths of 70 to 80 feet (21.5-24.5m) a fine silt appears. In general, sand is the principal substrate in this area, and masses of rock protrude above the sediment level like islands. Continuous rocky expanse occurs only centrally within the reef area.

Unfortunately the sands are relatively barren and desert-like. All the invertebrates characteristic of sandy bottoms are present, but they are typically sparse.

ROCKY BOTTOM

The richness of the rocky formations contrasts sharply with the impoverished condition of the sediments. At depths of about 20 feet (6m) where development of surf grass begins to thin, a dense and varied invertebrate turf appears, proliferating and thickening as depth increases. Composition of the turf was studied by Dr. Willis E. Pequegnat (see Bibliography) in the early 1960s. He found three different zones on the reefs, each with a fairly characteristic fauna. These were named reef top, midreef, and reef base. Invertebrate turf was best developed in the reef-top zone. Thick encrustrations also occurred in the midreef area, but at the bases turf was reduced to a patchy thin cover and motile scavengers such as urchins became common. The shoreward sides of the reefs were somewhat different from the seaward sides. Only 5 percent of the 300 to 400 species found by Dr. Pequegnat inhabited all three reef zones as well as the sandy areas between reefs.

The thickest turf rested on a basement of encrusting rock clams (*Chama, Pseudochama, Pododesmus*, and others), stacked on top of each other to form masses up to 16 inches (40.6cm) thick. When something frightens all the clams simultaneously, they close up in unison and the reef surface suddenly moves several inches. Dr. Pequegnat estimated that there were about 30,000 animals per square yard (square meter) in the thick portions of the turf.

Delicate coral-like bryozoans are common in the midreef zone. On the shoreward sides a pale orange species, *Diaporoecia*, is most abundant. Overall it is roughly cauliflower-shaped, but on a small scale is composed of tiny fingerlike projections with rough surfaces. On the seaward sides another bryozoan becomes common, the lovely orange-pink *Phidilopora*. The erect walls of this species are gracefully folded back on each other in convoluted clusters resembling large rosebuds. Many other colorful invertebrates contribute to these submarine gardens—anemones, corals, sponges, tunicates, to name a few groups. Several species of gorgonian corals proliferate on the larger pinnacles, giving the appearance of a delicate shrub cover.

Abalone and lobster are scarce—probably long since removed by humans. Whistler Reef is rich in fishes, however, although again pressure from fishing activity is evident. Large specimens are not common. The bright orange Garibaldi is extremely abundant. Garibaldi nests occur everywhere on the reefs during spring and summer. The parent clears away the invertebrate turf from a small patch of rock, and the eggs are deposited on this "nest." The parent guards the nest closely, chasing away other fishes and picking off encrusting invertebrates. A tiny filamentous red algae is allowed to grow, and the nest usually appears as a reddish patch on the rock. If you put a stone or a seashell on the patch, the parent will soon remove it. By fall all the eggs have developed, the nests are gone, and young Garibaldi appear everywhere. The juvenile is orange like the adult, but also has magnificent specks and streaks of irridescent blue. The breath-taking scenery of Whistler Reef becomes absolutely dazzling when large numbers of young Garibaldi are continually darting in and out of the crevices.

SUBMARINE FORESTS

Stands of Feather-boa and Giant Kelp grace the inner portions of Whistler Reef. For some unknown reason, kelp stops at depths of about 40 feet (12m) in this area. At these shallow depths, a substantial seaweed turf occurs here and there, particularly on the tops of pinnacles. Invertebrate turf still predominates on the sides and beneath overhangs. The kelp canopy is never thick. The bed probably is so shallow that pruning by storm prevents large masses of kelp tissue from accumulating at the surface.

4. THE HOOK (Fig. 18)

Location: The Hook (also called Fishhook Harbor) is a curving line of rocks barely breaking water on the east side of San Clemente Island. This reef lies about a mile from Pyramid Head, the southeastern tip. The rocks form a tiny harbor, and their semicircular arrangement gives the location its name. It is also called Fishhook Harbor. An excellent lee is afforded by the island against all weather except Santa Ana (easterly) winds. Even so, current directions change frequently, and it is unwise for vessels longer than about 25 feet (7.5m) to use the

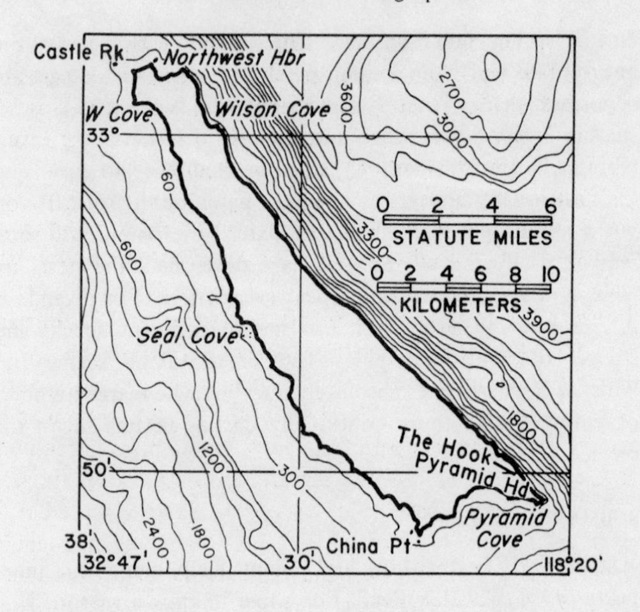

Fig. 18. Chart of San Clemente Island, showing location of The Hook (also called Fishhook Harbor). Bottom contours reveal the great escarpment lying along the east side of the island.

harbor. Excellent anchorage can be found immediately south of the rocks, but be sure to have plenty of anchor line (at least 200 feet [61m]) because the bottom drops off sharply.

San Clemente Island is 60 miles (97km) from the mainland. Nasty seas can build up in a few minutes, so don't attempt the crossing in boats designed only for coastal operation. For extra precaution have radio communication with the mainland. San Clemente Island lies off the beaten path, and if you are stalled, it may be a good while before a passing boat discovers you.

Services: The closest harbor facility is Wilson Cove at the north end of San Clemente Island. San Clemente Island is under the jurisdiction of the U.S. Navy, and Wilson Cove is a navy installation. It can be used only for emergencies. Avalon Harbor on Catalina Island is about 20 miles (32km) from The Hook. Restaurants, hotels, and diving air refills are available.

Access: The Hook is accessible only by boat. The shore is primarily steep cliff, so it is convenient to have a small skiff along for running from one spot to another when diving.

GENERAL DESCRIPTION

A majestic sea cliff borders several miles of the east side of San Clemente Island. The cliff top generally lies about 200 feet (61m) from shore at depths of 20 to 60 feet (6-18m). From there, vertical drops of 50 to 100 feet (15-30.5m) are common. The Hook is a massive bastion in this great rock wall. Here the sea cliff actually rises above sea level and plunges down to 180 feet (55m). The base of the sea cliff lies buried in coarse sand. The sand surface merely represents a diminution in slope, and the bottom rapidly drops away at about a 15-degree angle. Charts show very deep water lying close to shore along the sea cliff, hence there may be other undiscovered deeper cliffs down the sand slopes, below the scuba zone. Underwater visibilities here are about the best in all of California. Objects 100 feet (30.5m) away can usually be seen with ease. Most of the time the sea is beautifully calm. The diving is unexcelled.

POINTS OF INTEREST

SANDY ENVIRONMENT

Here and there sandy slopes extend into shallow water—generally where canyons from the island enter the sea. Sand slopes lie immediately north and south of The Hook. These slopes are quite barren in shallow water, and chunks of debris from rocky areas are about all that breaks the monotonous smooth surfaces. At about 80 feet (24.5m), sand-dwelling plants appear and animals become numerous. Debris and shells sliding downward meet these obstructions and are halted at least temporarily. This provides further substrate for plants and sessile animals normally attached to solid rock.

One of the dominant plants resembles a dwarf Elk Kelp. The large spherical float and great antlers characteristic of the genus are present but the stipe is usually less than 6 feet (2m) long–in contrast to the 80-foot (24.5m) stipes of Elk Kelp on the mainland. This dwarf Elk Kelp, *Pelagophycus giganteus*, is found only on San Clemente and Santa Catalina islands. The massive blades–up to 6 feet (2m) wide and 30 feet (9m) long–are draped in languid folds across the bottom. These are the deepest-growing plants recorded for California waters. Undoubtedly the exceptionally clear waters allow adequate sunlight to penetrate, thus supporting photosynthesis at great depths. Curiously, plants disappear from the adjacent rocky substrates at depths of about 100 feet (30.5m) and are replaced by invertebrate turf. The sand substrate, however, supports plant growth down to at least 200 feet (61m).

ROCKY BOTTOM

Shallow bottom lying within the sweeping circle of the harbor is crowded with one of the most colorful displays of shrub cover to be found anywhere. The dominant plant is the Southern Palm Kelp, *Eisenia*. Usually *Eisenia* is a dull brown, but in these clear island waters the hue becomes much lighter. At The Hook, palm kelp is golden-brown to yellowish with tinges of orange. The tendency toward lighter colors occurs in most other brown algae as well. A tropical seaweed, *Sargassum*, is common also (drifting *Sargassum* in the Atlantic Ocean gets trapped in a great whirlpool north of the Caribbean–the Sargasso Sea). Thin ramifications of the *Sargassum* blades are sometimes a shimmering blue underwater. Another thin brown alga, *Dictyota*, also displays blue hues. Both these plants are present at The Hook and create eye-catching blue patches scattered here and there among the rhythmically swaying fronds of yellow palm kelp.

Plants become sparse on the cliffs, particularly below 50 feet (15m), and encrusting invertebrates predominate. Several caves penetrate 10 to 20 feet (3-6m) into the rock formation. To avoid stirring up bottom sediments, swim along the roof when you enter a cave. The cave walls are lined with branching stony corals. Deep fissures and cracks at the backs of the caves are often loaded with lobster.

Occasional ledges provide a bit of horizontal surface on the cliff face and catch falling debris. These are good places to look for shells and fragments of coral. Plants often spring up on these ledges, creating little garden patches scattered across the cliff surface. At a depth of 80 feet (24.5m) a talus slope of large boulders has formed on a broad ledge, but this is not the bottom of the cliff. About 100 feet (30.5m) farther out, a massive series of pinnacles appears. Rocky bottom ultimately gives way to sand at about 180 feet (55m).

SUBMARINE FORESTS

Giant Kelp colonizes the sea cliff from depths of 20 feet (6m) down to about 100 feet (30.5m). Within the protection of the harbor at The Hook, Giant Kelp grows almost up into the intertidal zone. Like most other brown algae here, the island kelp is unusually light-colored. The blades grow to great lengths–some are close to 4 feet (1.5m) long and nearly 1 foot across. The Hook is one of the few places where the entire span of a 100-foot (30.5m) plant can be seen at once. All the fish schools associated with these great submarine trees can also be seen, hovering at various levels. It is worth noting, incidentally, that the fishes of the outer islands are much less afraid of man than individuals near the mainland. They do not hide as frequently, they approach more closely, and sometimes exhibit behavior patterns seen rarely, if at all, in waters where spearfishing is intense.

5. MORRO ROCK (Fig. 19)

Location: Morro Rock is a huge granite dome 576 feet (176m) high protecting the quiet harbor and village of Morro Bay from rampaging seas and furious waves that occasionally batter this part of the coast. Morro Bay lies 14 miles (22.5km) northwest of the junction of State Highway 1 and U.S. Highway 101 at San Luis Obispo, midway between Los Angeles and San Francisco. The region is only 65 miles (105km) north of Point Conception, so many of the relatively warm-water species

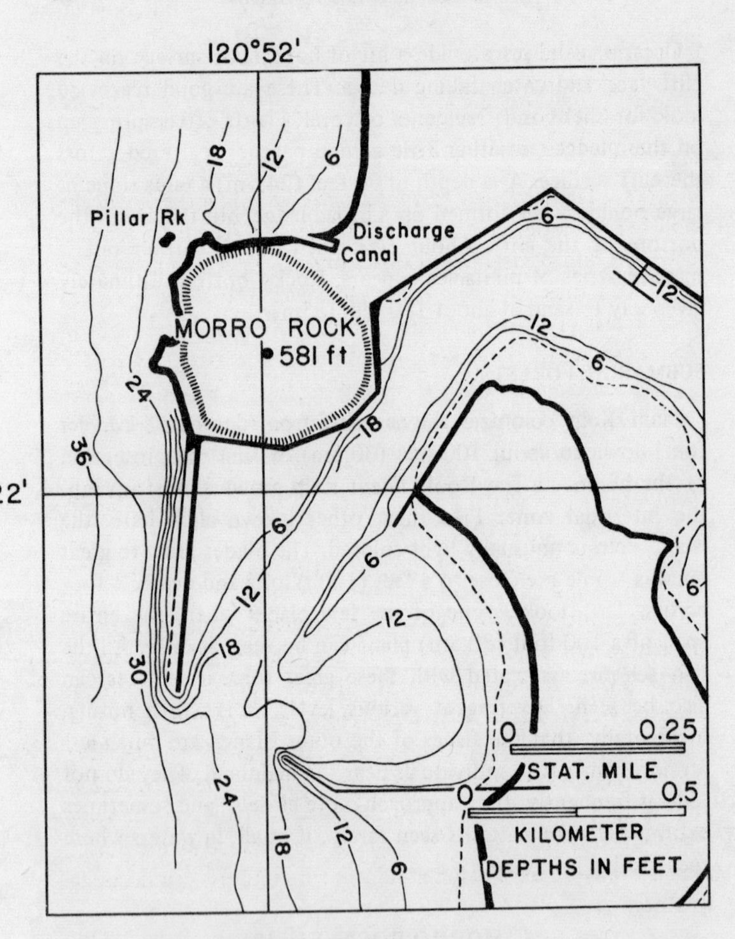

Fig. 19. Chart of western Morro Bay, showing Morro Rock and the discharge canal of the power plant.

occurring in southern California are represented in the local fauna and flora.

Services: Morro Bay derives income from two principal sources–fishing and tourists. The combination enhances the area for diving. Hotels, motels, and restaurants are easy to find and reasonably priced. A public boat-launching ramp is maintained, and privately operated hoist facilities are available. Vessels of nearly any size can be rented or chartered. A diving store in Cayucos, 5 miles (8km) away, provides compressed air, equipment rentals, and sales.

Access: The rock can be approached by land either from the sand beach to the north or from the jetty to the southwest. It is a short and pleasant boat ride from the docks to the outside part of the rock. Don't go diving, however, if the waves exceed 3 to 4 feet (1-1.5m) and don't use the jetty access except in very calm weather. Access from the beach to the north is slightly hindered by a canal discharging warm water from the nearby Pacific Gas and Electric power plant. Flows from the canal actually help in getting out to the Rock, but it is necessary to swim a wide arc to avoid the flow when returning.

GENERAL DESCRIPTION

The granitic sides of Morro Rock generally plunge steeply down to a flat sand bottom. Here and there small stretches of gentle slope occur, but most of the rocky substrate is essentially cliff. Large underwater crevasses penetrate the rock, and a few pinnacles protrude from the sand 50 to 100 feet (15-30.5m) away from the main rock. Depth where rock and sand adjoin varies from 20 to 30 feet (6-9m) along the west face (which is the most interesting place). Hence diving in this area is always shallow and should not be undertaken in rough weather.

POINTS OF INTEREST

SANDY ENVIRONMENT

Flat sand bottom lies west of Morro Rock. It is fairly desert-like, and sediment-dwelling organisms occur sparsely. Considerable debris can be found adrift on the bottom after storms, probably torn loose from the sedentary communities on the sides of Morro Rock.

ROCKY BOTTOM

The undersea slopes of Morro Rock support a shrub cover extending from the intertidal down to about 20 feet (6m). Below this level, as well as underneath the shrub cover itself, a luxurious turf mixture of seaweeds and invertebrates proliferates. A few feet above the sand-rock junction, attached life becomes sparse. In similar situations elsewhere, this barren band is usually scarcely noticeable. Perhaps the abrasive action of suspended particles is greater at Morro Rock because of the exposed location and ferocity of the seas. By the same token, sand levels may rise and fall over a greater vertical range and more frequently than normal, alternately burying and exposing the rock.

A substantial number of the invertebrate species at Morro Rock grow to unusual sizes. An urn-shaped sponge, *Rhabdodermella*, is usually less than half an inch (1.3cm) long. At Morro Rock, great packed clusters of individuals 1 or 2 inches (2.5 or 5.1cm) long produce a honeycomb effect on the undersides of overhangs. Many of the encrusting sponges and tunicates that ordinarily develop thin patchy layers, a quarter-inch (0.7cm) thick or so, occur here in ponderous slabs several feet across and 2 or 3 inches (5.1 or 7.6cm) thick. The Solitary Green Anemone, *Anthopleura xanthogrammica*, attains diameters of 1 foot (30.5cm) or more. Many other species are also uncommonly large here. The conditions that allow development of these monsters are completely unknown.

Some of the gentler slopes sustain meadows of lovely foliose red algae. Blood-red, lance-shaped blades of *Iridaea* 2 or 3 feet (61 or 91.4cm) long stream out in the swift bottom currents like pennants in a stiff breeze. Many delicate, almost transparent forms nestle in the protection of the densely packed turf. At mid-depths the relatively tall coralline alga, *Calliarthron*, in some places develops so abundantly that the bottom takes on the appearance of a pink grassland. Seaweeds do not attain the relatively great dimensions achieved by some of the invertebrates here, but their great abundance indicates that the environment is also unusually favorable for them.

The steepest slopes tend to occur along the southwest part of Morro Rock. More gentle slopes are found in the mid-region and in certain places along the northwest portion. Pillar Rock, a pinnacle projecting above sea level, lies off the northwest tip of Morro Rock and has both vertical and gentle slopes.

WARM-WATER CANAL

A large canal at the north base of Morro Rock discharges warm water from the Pacific Gas and Electric Company's Morro Bay plant. The water is taken in from the bay and heated about 16°F (10°C) within the plant before discharge. Diving in the canal is hazardous because of the fast flow, and access is prevented by a chain-link fence. The warm water disperses just north of Morro Rock, and large numbers of fishes are attracted to the area. The shore is always lined with fishermen, and on clear days this area is the best place locally for spearfishing. Farther north the sand beach is reputedly rich in pismo clams.

Diving within Morro Bay is often hampered by poor visibilities. A small bed of Giant Kelp grows opposite the discharge canal. Much of the bottom supports Eel Grass. Dock pilings and floats sustain excellent assemblages of invertebrates. Morro Bay is the southernmost enclosed body of water with a typical northern or cold-water species composition. The Plumose Anemone, *Metridium*, occurs abundantly on solid substrates, resembling a powder puff anchored to a thick stalk.

6. MONTEREY PENINSULA (Fig. 20)

Location: The term "Monterey Peninsula" is sometimes applied to the large area extending from around Moss Landing south to the Big Sur coast. We will use the name in a much more restricted sense and define the Monterey Peninsula as the region from the city of Monterey to the edge of Point Lobos. Monterey is reached either by driving up the spectacular coastal State Highway 1 from Morro Bay, or by leaving U.S.

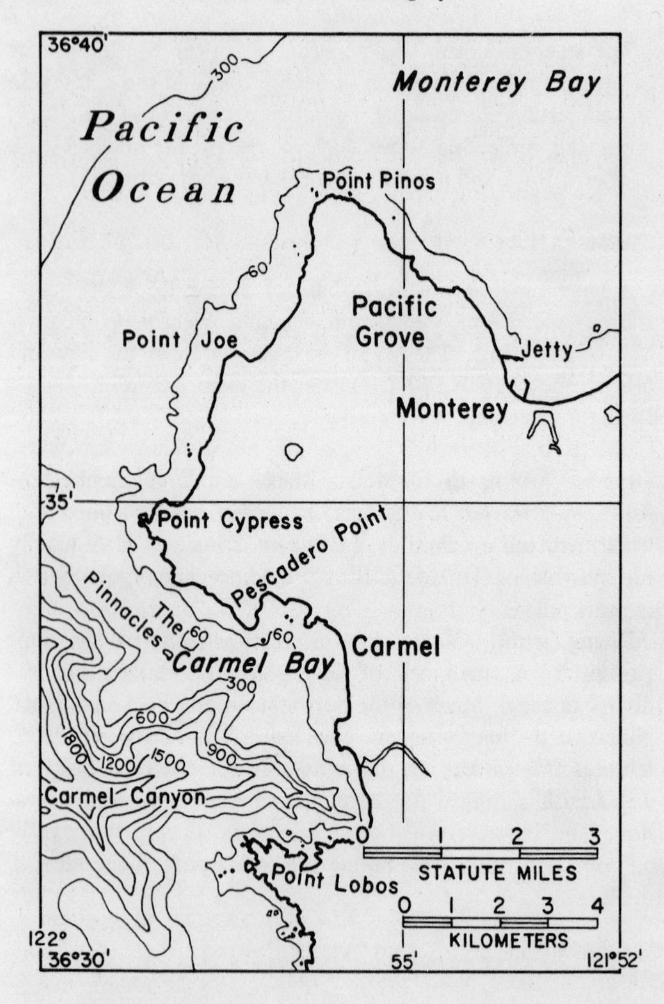

Fig. 20. Chart of the Monterey Peninsula, showing locations discussed in the text. Note the long extension of the Carmel Submarine Canyon that almost reaches shore at Mission Beach.

Highway 101 at Salinas, going west via State Highway 68. Monterey is also serviced by commercial airlines. The city is an important fishing port. Two nearby communities, Pacific Grove and Carmel, are primarily resort areas. A large, highly scenic portion of the peninsula, the Del Monte properties, is privately owned, but accessible on payment of an entrance fee.

Services: Hotels, motels, and restaurants abound. Fisherman's Wharf in Monterey is good for an infinite number of excellent seafood meals. Compressed air, diving equipment, rentals, and sales are available in Pacific Grove. Boat rentals are located on the municipal dock in Monterey harbor, and a special vessel catering to diving groups operates from the same area. A boat hoist for launching is also available on this pier.

Access: A large part of the shoreline is publicly owned or accessible after payment of a fee. The intertidal zone along Pacific Grove is a wildlife refuge where collection of boulders and animals is forbidden. Stanford University operates the Hopkins Marine Station at the east edge of Pacific Grove. The coastline from the Station to the beach at Lover's Point, and extending from the intertidal out 1000 feet (305m) offshore, is a preserve, and marine life may not be taken or disturbed. Large surf frequently hinders or prevents water entry west and south of Point Piños. Occasionally surf also limits diving east of Point Piños.

GENERAL DESCRIPTION

The Monterey Peninsula juts out between the principal branches of one of the world's largest submarine canyons. The continental shelf has a gentle slope off Monterey and Pacific Grove, and provides a broad expanse of relatively shallow bottom. The shelf narrows as the southern or Carmel branch of the Monterey Canyon is approached, west of Point Piños. Finally, along the southwest face of the Peninsula, the canyon frequently comes within a stone's throw from shore.

Fauna and flora of the Peninsula are almost entirely cold-water forms. A straggling few pelagic warm-water species may be seen occasionally. Likewise, warm-water benthic individuals may occur in a few semi-enclosed coves and downwelling regions that probably become slightly warmer than normal during certain seasons. In recent times the Peninsula has become populated with sea otters, which consumed huge quantities of urchins that had apparently dominated the ecology for many years. Disappearance of the urchins was followed by a massive development of vegetation, including dense submarine forests.

POINTS OF INTEREST

SANDY ENVIRONMENTS

Sandy beaches are generally an indication of sandy bottom offshore. Any kelp canopy offshore, on the other hand, is a most certain indication of rock substrate. The principal sandy areas are Monterey harbor and regions immediately north, the beaches directly off Carmel, and Mission Beach. Sand also occurs offshore from the kelp bed fringing Pacific Grove, and appears to be the principal environment surrounding the head of Monterey Canyon near Elkhorn Slough.

An unusual feature of the sandy bottom here is the occasional appearance of squid eggs. Adult squid, of course, are pelagic, but during reproduction the female buries one tip of her egg-case in the sand. The soft, flexuous, teardrop-shaped, translucent white case spends several weeks incubating before juvenile squid emerge. Frequently thousands of egg-cases occur packed together, covering large areas and resembling a snowfall. If you are very lucky, you may witness the squid mating and anchoring the egg-cases. At times they congregate in tremendous spawning schools and interfere seriously with underwater visibility. Squid are normally very shy and dart away when approached. During the mating period, however, they pay little attention to divers and can actually be handled gently. It is wise to wear gloves, since the animals can bite viciously and possess a venom. Squid occur along the entire coast, and the egg-cases are apt to be seen anywhere in deep water, but their favorite spawning places seem to be clean sand near submarine canyons where grain size is moderate. Judging from the frequency of egg-cases, the Monterey area must be the favorite spot for squid.

ROCKY BOTTOM

Most of the inshore shelf from Monterey to Carmel is rocky substrate. A few pocket beaches of sand interrupt the rock, but these are generally negligible. One significant sandy area in this long stretch is at Asilomar State Beach. Rocky sections also occur south of Carmel, but sand occurs more frequently. The topography is generally quite uneven, closely resembling the intertidal areas.

Plant communities attain maximum development here, perhaps from the influence of sea otters on large benthic grazers. Submarine forests and shrub covers are extremely dense. Underlying turf is usually a mixture of seaweeds and invertebrates. Foliose red algae in many places reach sufficient proportions to compete effectively with palm kelp and form a red shrub cover. Some of the common larger reds (*Iridophycus, Gigartina*) have a silver-blue irridescence when sunlight strikes at the proper angle. The fields of red streaked with fiery irridescence are a spectacular sight. Hosts of small invertebrates inhabit upper portions of the shrubbery, but they are usually well camouflaged and hard to find. Some of the brilliantly colored turbans and top snails, however, stand out conspicuously.

The underlying turf is generally crammed with invertebrates. Collection is often impeded by the thick shrub cover. Sometimes the only way to recover turf is to uproot the entire shrubbery with the turf attached and dissect it topside. Invertebrate turf becomes easily accessible beneath overhangs and under flat stones. The shrub cover of cold-water climates may be somber-hued compared to warm-water cover, but the spectacular coloration of encrusting invertebrates in the turf more than compensates. The reds and oranges are subdued in the blue submarine light, but can be appreciated with a flashlight. If you are collecting small invertebrates concealed in the turf, it is often wise to use a flashlight even when illumination by sunlight is good. In the natural blue light the coloration of many animals frequently camouflages them, but they stand out conspicuously in the beam of artificial light.

Abalone, urchins, and larger invertebrates are scarce. They must be sought in crevices and under very large rocks that sea otters cannot overturn.

SUBMARINE FORESTS

Three kelps of the Monterey Peninsula grow sufficiently large to form a forest environment (*Cystoseira, Macrocystis*, and *Nereocystis*). Giant Kelp (*Macrocystis*) requires some protection and does not grow extensively along the most exposed part of the coast from Point Piños to Cypress Point. *Cystoseira* is also reduced and pruned back in this exposed region. Only Bull Kelp (*Nereocystis*) can develop sufficiently dense colonies here to create forests. As the environment becomes calmer and more bay-like near Monterey, however, Bull Kelp thins out and the other two species dominate. The vegetative part of *Cystoseira* is small and shrub-like. The reproductive foliage forms the main mass, developing into a dense thicket that is difficult for divers to penetrate.

In the thicker parts of the *Macrocystis* beds, canopies shade the bottom so effectively that small-statured plants disappear completely and the bottom becomes covered with pure invertebrate turf. Translucent masses formed by colonial tunicates are often the dominant encrusters in cold waters. They are usually colored in pastel shades, and form surrealistic shapes sprawling across the bottom.

SUBMARINE CANYON

The Monterey Submarine Canyon is divided into two principal branches. The Monterey Branch lies in Monterey Bay, approaching shore near Elkhorn Slough. It has not been explored in detail, but is primarily lined with soft sediments in its shallower reaches. The Carmel Branch of the canyon extends into Carmel Bay between the Monterey Peninsula and

Point Lobos. It is more easily accessible than the Monterey Branch, and has some fine rocky areas.

The Carmel Branch borders the shoreline of the Monterey Peninsula from Cypress Point to Arrowhead Point. Access from the beach is hindered in most places by very dense kelp canopies. The best way to work in this area is by diving from a boat. If you do not have a fathometer to locate the edge of the canyon, dive at the outermost edge of the kelp bed. In many places, stands of Bull Kelp grow right at the upper borders of the canyon; the region off Pescadero Point is one such location. Poor beach access has reduced diving activity in this part of the canyon, and the environment is close to its native condition. Don't attempt to dive here if you are a novice, because the upper parts of the canyon lie at about 100 feet (30.5m).

The shallowest and most easily accessible portion of the Carmel Canyon lies off Mission Beach. A small kelp bed runs out from the north end of the beach. Near the outer edge of the bed, about 400 feet (122m) from shore, a reef breaks the surface at low water. The reef marks the edge of the canyon. Swim out along the border of the kelp bed until you are parallel with the reef. Descend here (depth 40 feet [12m]), and the edge of the canyon will be about 50 feet (15m) farther ahead.

This spot is interesting because the edge of the kelp bed lies at the junction of sand and rocky environments. It is one of the few spots where sand along the edge of a canyon supports resident organisms. Plants are able to grow on small solid objects embedded in the sand. This may enrich the area and add some stability to the sand.

The rock bottom slopes gently downward at first, but rapidly steepens. One of the first things that catches the eye are spectacular giant anemones of the genus *Tealia*. The columns vary from red to snow-white, and the disk and tentacles are streaked with combinations of red, orange, yellow, and white. They resemble huge flowers and stand out vividly in the deep shade beneath the kelp canopy.

The region of the rock-sand junction tends to collect debris that somewhat modifies the area. A few yards north, however, the rocky communities attain an extraordinarily rich development. The most luxurious region here is associated with the edge of the kelp bed, roughly between depths of 70 and 90 feet (21.5-27.5m). Shrub cover and seaweed turf appear where shading from the kelp canopy lessens. Plant life extends below 100 feet (30.5m) here but becomes quite sparse at such depths, probably because of reduced light intensity. For those interested in unusual red algae, this zone at the edge of the kelp invariably yields fine collections. For some reason, it is continually changing. You can return after a few months and find a completely new flora.

Within the kelp bed, shading reduces or eliminates associated vegetation, and the invertebrate turf proliferates. Like the communities at Morro Rock, many species tend to become unusually large. In some cases distributions are slightly modified. For example, a blue sponge that ordinarily forms thin encrustations far back under overhangs (*Hymenamphiastra*) actually occurs on the sides and tops of the huge boulders here. It forms rich velvet carpets colored a deep cobalt.

Below 100 feet (30.5m) both the seaweed and the invertebrate turfs are reduced. Large boulders become scarcer, and steeply inclined rock faces dominate. The slopes remain precipitous to depths far below the scuba zone, so don't try to reach level bottom here.

POINT LOBOS AND SEA OTTERS

Point Lobos State Reserve begins at the south end of Mission Beach. It also borders the south side of the Carmel Canyon, but is not a good access area for canyon diving. The shelf is generally broad between the beach and the canyon, and thick kelp beds intervene in most places. The Reserve is very interesting in its own right. Collecting is strictly prohibited, and a completely natural situation has been preserved. Diving is permitted but is strictly regulated. Only a few divers are permitted in the water at the same time, and diving groups must contain at least three divers or two divers plus a shore observer. Permits for diving are issued at the entrance.

The Monterey Peninsula is one of the few places where Sea Otters can be observed. Underwater encounters were formerly rare, because the animals are very shy, but recently some Otters have learned to accept squid and urchins from divers. They are easily seen from shore, particularly in early morning when they are foraging. Lovers Point in Pacific Grove is a fine area for Otter-watching.

7. SALT POINT (Fig. 21)

Location: Salt Point lies just south of the Kruse Rhododendron Reserve State Park, about 10 miles (16km) north of Fort Ross. Turn off U.S. Highway 101 at Santa Rosa. Go 7 miles (11.5km) to Sebastopol, and proceed to the coast at Bodega Bay (14 miles [23km]) or by State Highway 12 to Jenner (25 miles [40.5km]). Go north on State Highway 1 to Salt Point (20 miles [32.5km] from Bodega Bay, 11 miles [18km] from Jenner). The property surrounding Salt Point has recently been acquired by the State Department of Parks and Recreation, and is being converted to a state park.

Services: Services at Salt Point have been virtually nonexistent, but will become available as the state park is developed. It is likely that the nearshore waters will become one of the state's first underwater parks. Probably camping and picnic facilities will be constructed. Other facilities will undoubtedly be installed, and perhaps dressing rooms for divers will be provided.

Access: The principal access area is Gerstle Cove, just east of and in the lee of Salt Point. Protection from surf is excellent, and water entries are nearly always feasible. A steep road leads down to the beach at the west end, and small portable boats can be launched. More difficult access is possible down the cliffs at the east end of Gerstle Cove and just west of Salt Point. These areas, however, are more exposed. Undoubtedly access will be improved as the state park is developed.

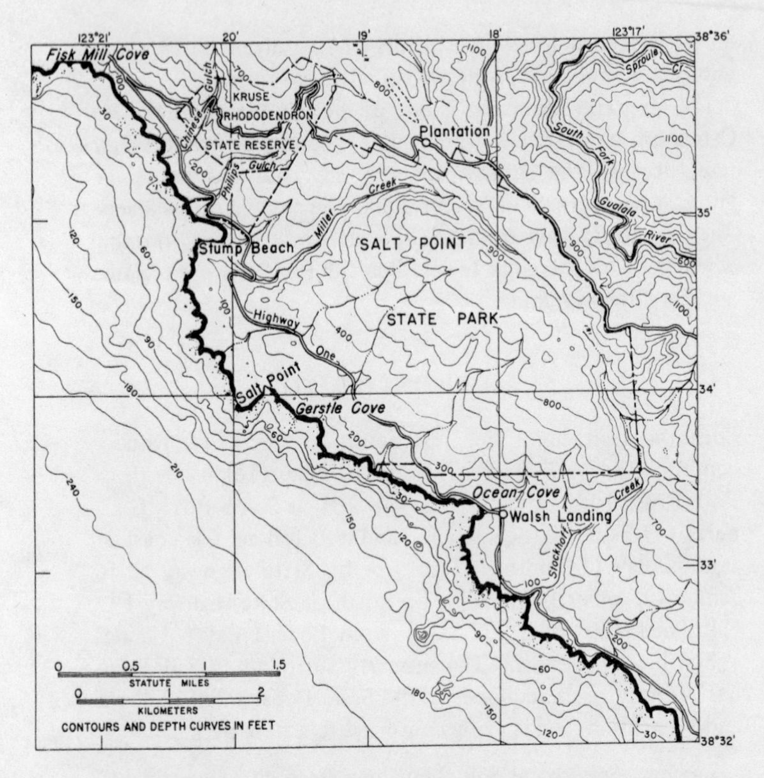

Fig. 21. Chart of the coast surrounding Salt Point, showing the State Park and the underwater park that lies offshore.

GENERAL DESCRIPTION

The Salt Point environment is primarily a rocky bottom. Like most of northern California, the coast is exposed to severe weather that strongly influences the fauna and flora. Although Gerstle Cove is well protected, the benthic communities tend toward associations characteristic of exposed coast. Just south of Gerstle Cove, pieces of wreckage are scattered across the bottom. These are the remains of the 385-foot (117.5m) freighter *Norlina*, which ran aground in heavy fog on August 4, 1926.

POINTS OF INTEREST

ROCKY BOTTOM

Except for a small pocket beach at the north end of Salt Point State Park (Stump Beach), the environment consists entirely of rocky bottom. A few patches of Bull Kelp are scattered along the coast, but none constitute a dominating stand that might be termed a submarine forest. Nonetheless, several community types occur.

The shallowest zone is primarily short seaweeds that are not sufficiently tall to be considered shrub cover, yet usually not so dense as to form a good turf. Most are foliose and harbor a substantial fauna of motile invertebrates.

At depths of 5 to 10 feet (1.5-3m) a very dense shrub cover develops. The dominant plant is the Northern Palm Kelp, *Pterygophera*, growing to the unusual size of 6 feet (2m) or more. A mixed turf carpets the bottom with a rich assortment of plants and animals. Large smooth boulders are common, and legal-sized red abalone can be found by the sharp-eyed. Here and there dense patches of urchins dominate partially or completely. Salt Point is presently beyond the northern limit of Sea Otters in California, and urchins plague much of the nearshore bottom.

The bottom slopes gently seaward and is composed of flat slabs of basement rock interspersed with boulder patches. Here and there the basement rock erupts into cliffs and pinnacles. The larger pinnacles are often marked by patches of Bull Kelp. They are excellent diving spots. The shrub cover of Northern Palm Kelp remains dense, but the plants are much shorter than in the shallow areas. Animals tend to dominate the turf, but many small plants nonetheless occur. The Pinto Abalone (*Haliotis kamschatkana*) occurs in abundance in many places. The Gumboot (*Cryptochiton*), a huge chiton that looks like an upside-down brown shoe lying on the rock, is easy to find. Very large octopus are seen occasionally in the crevices, and many of the cliffs are adorned with clusters of large white anemones (*Metridium*). These creatures have columns up to 1 foot (30.5cm) long and are topped by masses of tiny tentacles grouped in lacy folds, giving the overall appearance of a powder puff on a stalk. The brilliant whiteness of these creatures contrasts strikingly with the dull gray background of the cliffs and enhances the loveliness of the delicate crown of tentacles.

At depths of about 50 feet (15m), life becomes sparse except for large urchins. These animals dominate the huge smooth boulders that are scattered in disarray everywhere. The region obviously needs a band of Sea Otters to restore the balance of nature.

8. MENDOCINO COAST (Fig. 22)

Location: The town of Mendocino is well off the beaten path, ten miles (16.1km) south of Fort Bragg and about 130 miles (210km) north of San Francisco. Like many coastal spots of northern California, Mendocino straddles State Highway 1 and requires a journey west from the main road, U.S. Highway 101. The principal roads to Mendocino are State Highway 128 from near Cloverdale when coming from the south (55 miles [88.5km] to the coast), or State Highway 20 from Willits when approaching from the north (37 miles [59.5km] to the coast). The coast is rugged and broken with many islets, pinnacles, and reefs. A 17-mile (27.5km) stretch in the vicinity of Mendocino offers many fine places to dive.

Services: Motels and restaurants are available in the principal communities of Albion, Mendocino, and Noyo-Fort Bragg. Mendocino has a diving store with equipment, air supply, and rentals. Boat rentals and launching facilities can be obtained in Albion and Noyo. Camping is allowed in Russian Gulch State Park and at Van Damme Beach State Park.

Access: Private ownership hinders access in many places. Nonetheless, sufficient public lands with suitable beaches exist to give a good sampling of the various parts of the Mendocino coast without resorting to a boat. Especially noteworthy are Russian Gulch State Park (not to be confused with another

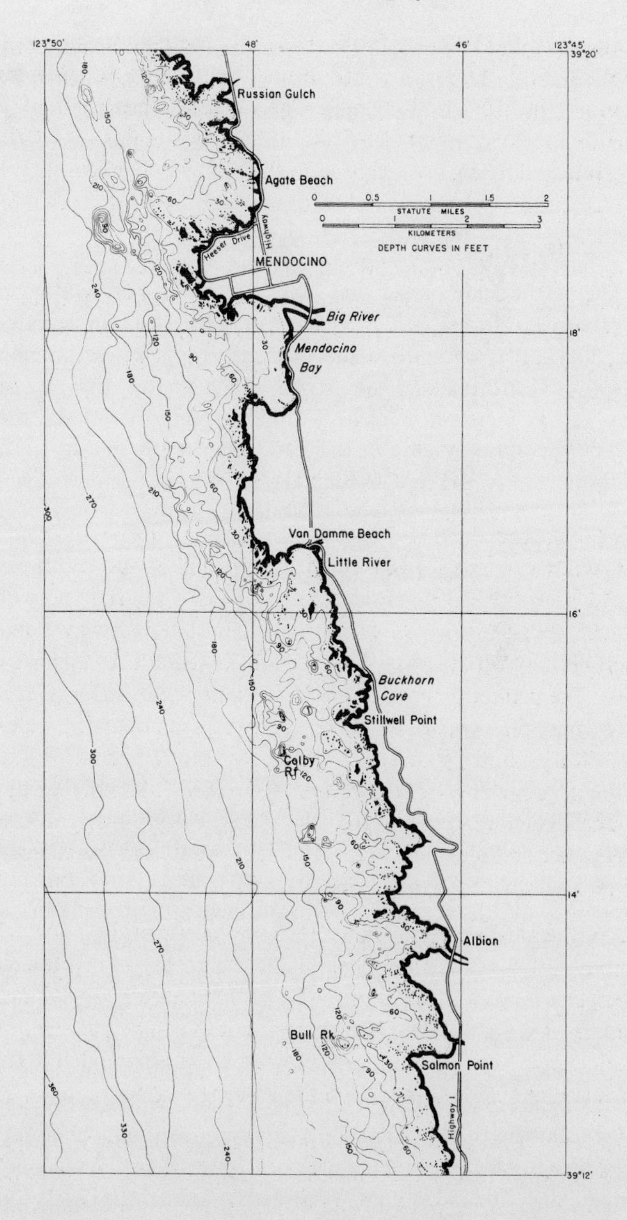

Fig. 22. Chart of the coast from Russian Gulch to Salmon Point, showing locations discussed in the text.

Russian Gulch in Sonoma County), Van Damme Beach State Park, and the bluff area just west of Mendocino, reached by Heeser Drive. Beach parking is available at Van Damme Beach, but short hikes down cliffs via trails are usually necessary elsewhere.

GENERAL DESCRIPTION

The Mendocino coast has about the clearest water in northern California (which is not noted for good underwater visibility). It is a luxuriant region, and many species common north of California have their southern limits along this part of the coast. Substantial coastal shipping activity served the lumbering industry here in the latter nineteenth century. The harbors and coves are vulnerable to storms, however, and remains of wrecks are common. Nonetheless, there is sufficient lee to allow growth of Giant Kelp, and this area is one of the few spots in northern California where large *Macrocystis* beds occur.

POINTS OF INTEREST

SANDY ENVIRONMENT

The protected embayment off Van Damme Beach displays large sandy patches dotted with cobbles and boulders. Gravel beds also occur. Like most northern areas, the waters are apparently loaded with plant nutrients; algae grow on any attachment surface available–worm tubes, bits of shell, etc. A goodly amount of drifting seaweeds seem to do well here in the protected waters. The general abundance of plant life in the sandy areas modifies the ecology slightly; fishes normally associated with rocky bottom roam out into the sedimentary territories here. The abundance of drifting algae encourages certain grazers that depend on such food. Noteworthy are large numbers of Red Abalone. Large specimens may be found attached to small boulders not much larger than the abalone, totally surrounded by sand. Possibly the animal spends its life on a single small rock, nourished by the abundance of drifting seaweeds.

ROCKY BOTTOM

The Mendocino coast excels in rocky bottom that is twisted and sculptured into every conceivable shape. Underwater cliffs, pinnacles, caves, arches, ledges, tunnels, etc., abound from the intertidal out to about 40-foot (12m) depths. Good diving continues even deeper, and abalone occur in abundance out to depths of at least 80 feet (24.5m). The scourge of rocky bottoms, urchin hordes, also dominates over substantial territory. Seaweeds develop in surge channels and other areas where wave action prohibits intense urchin activity. Bull Kelp forms patches of forest environment, and shrub cover is provided by Palm Kelp and large specimens of Blade Kelp.

Truly incredible invertebrate turf encrusts the cliffs and undersides of ledges. Frequently such turf is rather drab in appearance, due to dust-like sediments adhering to mucous surfaces of the animals. The invertebrate turf here, however, tends to be extremely colorful. Possibly the dust-sized particles are rare, or the strong surge may cleanse and scour the sticky surfaces.

Chapter 6

AN INTRODUCTION TO UNDERWATER PHOTOGRAPHY

by Robert Hollis and Wheeler North

"Specimen collecting" with a camera is an exciting way to show the beauty of the undersea world to those who will never see it for themselves. Photographs display the living marine inhabitants in their natural settings, which dead specimens can never do. The photographer also has the satisfaction of knowing that he or she has not destroyed the delicate ecological balance that is so important to maintaining the abundant life of the sea.

This discussion will assume that you have a basic understanding of photographic principles. (If you do not, then learn to take good pictures on land first, because it is cheaper in time and materials than underwater.) The submarine environment is more complex and changeable than the terrestrial medium, and it creates special problems for the aspiring photographer. Knowledge of a few elementary concepts will help you quickly evaluate the great variety of situations you may encounter.

What follows is a brief explanation of the physical principles you must know and a survey of equipment and techniques. For detailed advice on cameras, housings, lighting, film, etc., go to a diving equipment store with a good photographic department.

LIGHT AND COLOR UNDERWATER

Sea water is about 1000 times as dense as air. The influence of the watery medium on the passage of light increases in roughly the same proportion as the density: effects which the terrestrial photographer can ignore become multiplied a thousandfold underwater and must be considered. Correctly evaluating a situation requires that you understand a few basic physical laws that govern light in the ocean.

ABSORPTION AND SCATTERING

Two principal factors affect light underwater: absorption and scattering. *Absorption* is reduction of light intensity by the water itself. *Scattering* is a change in the direction of a tiny light beam caused by reflection from a suspended particle. Each affects underwater visibility in different ways.

Suppose you are looking at an object quite a few feet away in very clear water that has no suspended particles (i.e., no scattering). As you move away from the object it appears dimmer, because more of the light coming from it is *absorbed* by intervening water. Finally you can see the object clearly only if it contrasts strongly with the background (i.e., if it is much brighter or much darker). The image quality, however, remains unchanged–that is, light from one part of the object has been reduced just as much as light from another part. If you could look at the object through an underwater telescope, fine details could still be seen.

Now suppose you are looking at the object in a strongly *scattering* medium–one containing many suspended particles. As distance from the object increases, fine details are lost (Fig. 23). Light from one part is not the same as light from another part because of the random nature of scattering. The quality of the image degrades. The object may contrast strongly with its background and be seen easily, but details are hazy. Even in apparently clear waters, haziness effects become noticeable when subjects are 15 to 20 feet (4.5-6m) away.

Think of your camera as a poor human eye–it needs more intensity than the eye to record an image. It adapts to changing conditions only if you alter the settings properly. It cannot integrate a hazy image in time to make out blurred details. To compensate, you should photograph only subjects that you can see extremely well. Hazy objects make good backgrounds, but the principal subject should be clear and well

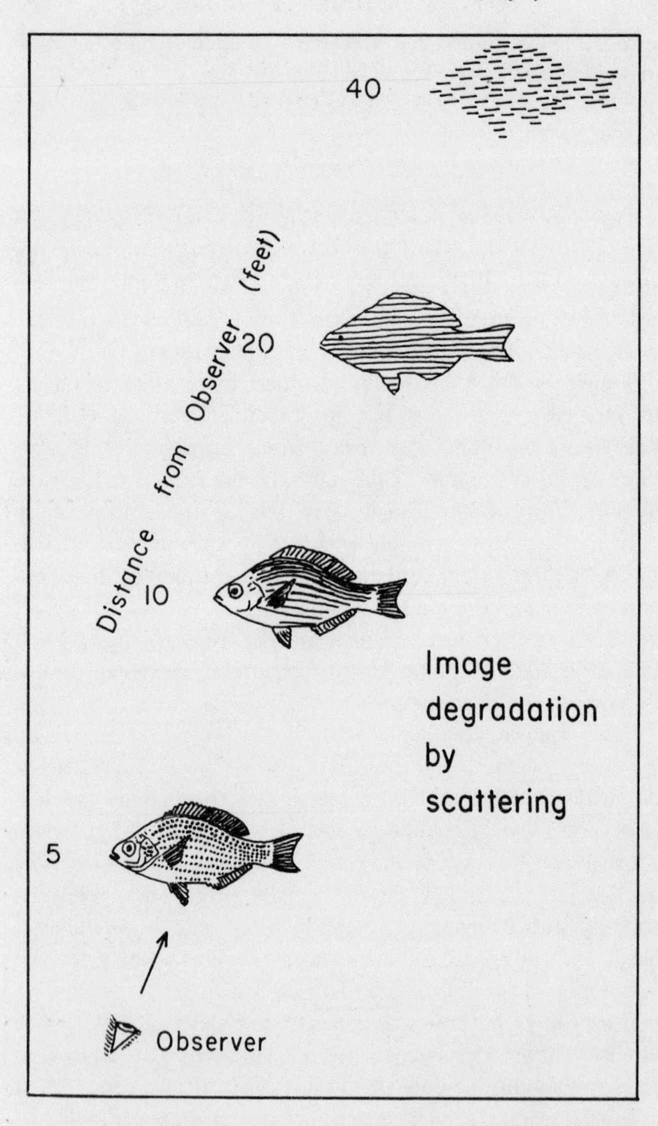

Fig. 23. Diagram illustrating degradation in image quality underwater as distance from observer to object increases. This type of degradation results from the scattering of light rays by tiny suspended particles.

illuminated. A general rule is to avoid shooting subjects that are farther than one-fourth the distance you can see.

REFLECTION

A portion of the sunlight entering the sea is *reflected* from the surface of the water, so submarine light intensity is almost always lower than topside values. The amount of light reflected varies, depending both on (1) the angle of the sun and (2) the roughness of the sea. However, the sun's angle is far more critical as it changes throughout the day. Figure 24 shows the percentage of light lost by reflection at different times of day. Note that light losses are greatest at low sun

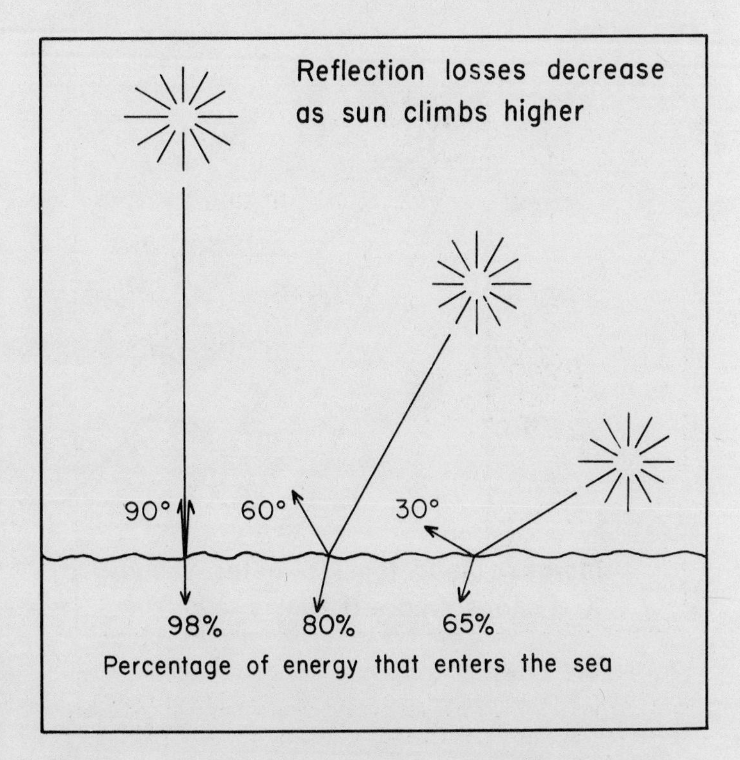

Fig. 24. Diagram showing how light entering the sea increases as the sun's position rises in the sky.

angles (early morning or late afternoon) and least when the sun is overhead.

Because of the continually changing amount of light penetrating to the submarine world, an underwater light meter is considered essential by most photographers to determine the proper f-stop setting for their cameras. However, if you have only a land-based light meter and want to do some experimenting as a beginning underwater photographer, you can roughly estimate the proper f-stop settings by using the information in Figures 24 and 25.

Underwater photography is best attempted under the following conditions:

1. When the sun is overhead, between 10 a.m. and 2 p.m.

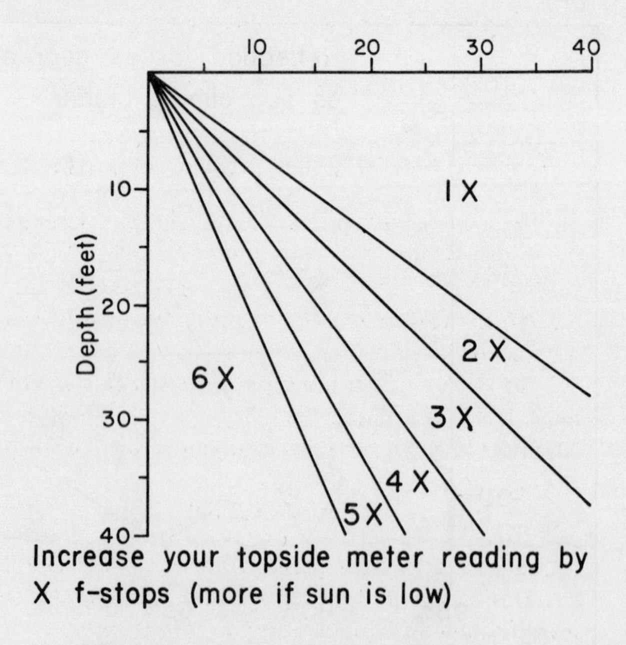

Fig. 25. Graph for developing estimate of proper f-stop to use underwater if you have a topside light-meter reading. Depending on underwater visibility and water depth, the graph predicts how many f-stops you must open up the diaphragm over the f-stop indicated by the light meter.

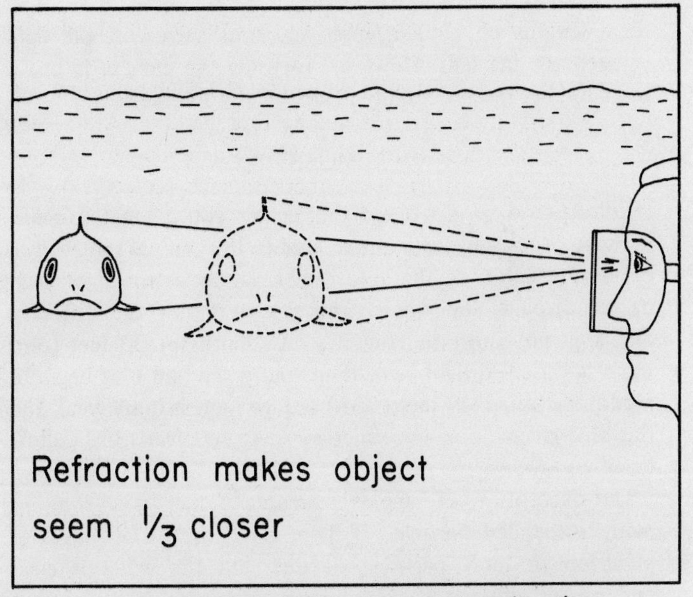

Limit of Visibility Underwater (feet)

Fig. 26. Refraction of light rays entering the diver's facemask creates an apparent image that is about one-third closer than the actual object.

2. When weather is calm, so reflection of light caused by wave action is at a minimum.

3. In shallow water, because the distance the sun's light must travel is reduced, and because it is possible to take advantage of any light reflected upwards from the bottom.

REFRACTION

Light rays bend when passing from air to water and water to air. This change in direction is called *refraction*. It creates a magnification effect underwater, so objects appear to be about one-third closer and somewhat larger than they really are—to both your eye and your camera (Fig. 26). Refraction occurs when (1) the light enters your facemask and (2) when it penetrates the housing of your camera, slightly reducing the focal lengths of camera lenses and thus narrowing the field covered by the lens. However, focusing the camera is not a problem, because essentially it sees what your eye sees.

COLOR

Colors tend to disappear the deeper you go in the ocean because of *selective absorption*, that is, the gradual subtraction of all the colors in the spectrum from the white light as it passes through the water. The absorption of the different colors is not uniform (Fig. 27). At depths of 20 feet (6m) there is little red light remaining, and green and blue begin to predominate. At 40 feet (12m) orange light is quite weak and the blue-green dominance increases. At 100 feet (3m) yellow becomes low.

The absorption of blue is increased if the water contains many suspended particles. If there is a dense surface layer of plankton or inert particles of dirt, but the water is clear underneath, green will predominate. However, if the surface particles are large, some of the green may also be absorbed, resulting in a prevailing yellow-green hue.

These color shifts alter the appearance of many objects underwater. A red-orange fish like the garibaldi, for example, appears orange at 10-foot (3m) depths, yellow-orange at 20 feet (6m), and yellow at 30 feet (9m). Pure red objects, like some sponges, will look quite drab at depths greater than 30 feet (9m). Artificial light is required to bring out the true colors at the greater depths.

Films that are primarily sensitive to red light are not of much use below 10 feet (3m). The color shifts can be compensated for to some extent with filters in shallow water. But below 20 feet (6m) so little natural red light remains that adequate compensation would cut the total intensity below acceptable levels.

Bright red, orange, or yellow organisms are occasionally seen at depths greater than 30 feet (9m). These creatures may not be reflecting the small amounts of colors available in the ambient light–instead they may be fluorescing. *Fluorescence*

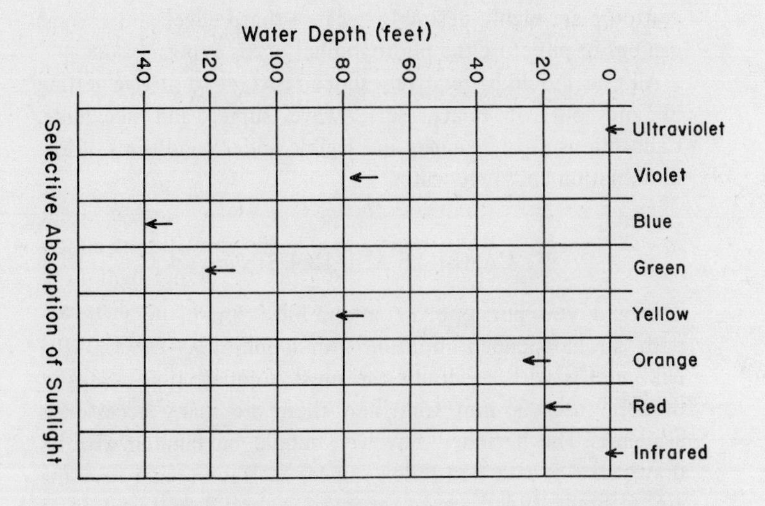

Fig. 27. Diagram showing relative strengths of different colors of light underwater, in terms of the greatest depth at which a color would contribute significantly to an underwater photograph, under clear-water conditions (visibility greater than 30 feet). As water turbidity increases all depth figures would decrease, but violet, blue, and green would be affected most strongly.

is accomplished by absorbing light of one color (in this case blue) and emitting another color (red, orange, or yellow). Although such organisms are eye-catching and stand out in their environments, the intensities they emit are weak. Fast films and wide diaphragm openings are necessary to photograph them.

TIPS ON EQUIPMENT AND MATERIALS

A beginner in underwater photography should keep equipment to a minimum and avoid complexities. Because of the fine parts and movements inherent in most cameras, damage from corrosion and water leakage must be guarded against carefully and continuously. Equipment should sink when released, but not be heavy and burdensome. Small size and streamlined contours are highly desirable because sharp edges and corners can cut or puncture the photographer or his exposure suit.

All gear should be relatively shock resistant to survive getting in and out of boats, surf, wave surge, and accidents. Calibrations must be clear and legible under conditions of low illumination and dirty water.

CAMERAS AND HOUSINGS

Should you purchase an amphibious camera or a camera with an independent housing? An amphibious camera (the Nikonos) is without doubt the most popular. It is compact, durable, uses 35-mm film, and there are many accessories available. The beginner, however, should be familiar with its drawbacks–it is a non-reflex camera so framing and focusing are difficult and done by estimation, and lens selection is limited. However, it is an excellent camera to start with, as it is the smallest unit available of comparable quality and capability.

You may decide to purchase an independent watertight housing for a camera you presently own, or you may wish to purchase a camera and have a housing constructed for it. Remember that there will be maintenance requirements for both the camera and the housing. Housings sometimes need more maintenance than cameras, and they can cost as much as or more than the camera. Nonetheless, an efficient well-designed housing allows you to take a greater variety of good pictures because it incorporates many accessories that provide the different effects you may want.

Before purchasing a watertight housing, check the following points: Are there enough controls (f-stops, focus, etc.) available for your needs? Can housing be opened easily for film changing in the field? Is buoyancy satisfactory (slightly heavy with camera inside)? Can the subject be viewed through the camera when you have a face mask on? What is the maximum depth of use? Can water leakage become a problem? (O-ring seals usually give more trouble-free performance than gaskets, but proper maintenance of the latter can eliminate any cause for worry.)

Plexiglass housings are lightweight, transparent, and usually cost less than metal housings. However, they are positively buoyant and considerably more fragile. Anodized aluminum housings do not require buoyancy-compensating weights, but they are of course heavier out of water. Although metal housings are initially more expensive they retain their value and incorporate finer features than the plastic ones.

Numerous accessories are available, and many are imperative in achieving high-quality photographs. When you purchase your underwater camera and/or housing, be sure that accessories can be added. Don't get locked into one type of photograph.

LIGHT METERS

The light-sensing element in light meters can be selenium or cadmium sulfide. Cadmium sulfide (Cds) types are usually more expensive but have greater sensitivity, which is needed in the dimly illuminated underwater world. If your camera is not equipped with a coupled exposure meter, you will need a separate photoelectric light meter. It should be of the reflectant type, rather than incident. The incident meter is of limited use underwater, and then only in shallow water. Securing your meter to the camera housing by a bracket helps to avoid damage from jarring.

Your light meter should be provided with a commercial watertight housing. If your budget does not permit this, you can improvise a relatively satisfactory waterproof housing for a light meter from a glass mason jar with a rubber seal in the lid. This works fairly well in shallow water, but be sure to test it for pressure before taking it to any depth.

LENSES AND VIEWERS

A camera with a fast lense is a necessity: openings of f-3.5 or wider are constantly required underwater. Wide-angle lenses are preferred to compensate for the refraction and image-degradation problems discussed above.

The usual camera viewfinders are difficult or impossible to sight through when underwater, because the mask keeps your eyes too far away from the viewfinder port. Consequently, sports finders and prism-reflex viewers are commonly used in underwater photography. *Sports finders* are devices that mount on the camera top and frame the viewing field. There is usually a parallax problem—the line of sight of the diver differs from that of the camera. Furthermore, the sports finder offers no aid in focusing. The *prism-reflex sports finder* provides through-the-lens viewing. Mirrors redirect the view of the camera into the eye of the diver. Single-lens reflex cameras equipped with prism-reflex sports finders provide an image free from parallax; the subject is viewed directly through the shooting lens. Twin-reflex cameras sometimes introduce parallax when there is not sufficient compensation for line-of-sight differences between the viewing and focusing lens and the shooting lens.

Dirty lenses, which might yield acceptable topside pictures, can shift the quality from good to bad in a moderately scattering watery medium. Rinse equipment thoroughly in fresh water after immersion in salt water; take flash units completely apart and operate all controls while rinsing. During a dive, check housings periodically for leakage.

STILL SHOTS

In choosing subjects for underwater stills, remember that much of the fascination in diving derives from movement and suspension in space; this feeling is often lost in still photography. The principal subject should be fairly small. Massive coral pinnacles or steep impressive canyons usually turn out as hazy ill-defined shapes in photographs. An interesting shot will show the entire subject in clear focus and with reasonable contrast. Larger objects such as the walls of a canyon or the ribs of a ship can, however, be used effectively as a background.

Beginners will find it less expensive to start with black-and-white film. Maintain a steady position while photographing; this will probably require bracing yourself and exhaling before shooting. Sometimes it is desirable to add extra weights or pick up stones. Avoid stirring up bottom sediments beforehand. Photograph other divers as they exhale—bubbles add interest to a picture. A background of fishes almost always increases the artistic quality so it is usually worthwhile to wait for a nearby school to swim into the needed location. Fishes can often be attracted by scattering chum, such as crushed sea urchins or other invertebrates.

Avoid shooting straight down; this tends to produce uninteresting effects. Downward angles, horizontals, or verticals are more apt to be good. Blurring is always a problem because of the close working distances and slow exposure times. For moving objects, head-on views are less likely to be blurred.

Deep water photography, although posing greater lighting problems, is often easier than shallow work because of reduced water movements. Visibility may also be improved.

CONTRAST

Getting good contrast is nearly always a problem in underwater photography. Solid nonliving surfaces in the sea become coated with a thin scum that tends to cause everything to reflect light more or less uniformly. Before taking each picture, ask yourself "How can I increase the contrast?" Perhaps shooting from a different angle will help. Perhaps the subject can be moved to a more contrasting background. Shooting from low angles often improves contrast by providing a sharp silhouette along the upper edge of the subject (Fig. 28). Other divers in their dark exposure suits are always difficult subjects, because large black areas detract from a picture; very light background will compensate somewhat for the blackness of the suit; sometimes it is best to show only the diver's arm with a spear, or a hand with a collecting tool.

Contrast achieved by silhouettes; improves visibility

Fig. 28. Diagram showing usefulness of silhouettes for photographing objects that do not contrast strongly with their background.

ESTIMATING EXPOSURES USING NATURAL LIGHT

Before taking a picture underwater, using natural light, establish the environmental lighting by a light-meter reading (or by estimation or computation from Figs. 24 and 25). Tilt the meter slightly downward, or shield it from light from above by the palm of your hand. Then evaluate the subject. If the subject is lighter than or about the same shade as the background, use the f-stop given by the light meter reading. If the subject is darker, open the diaphragm one or two stops. Of course a dark subject should be framed against a light background, and vice versa, for contrast if possible.

Many underwater objects are so dark that it is difficult or impossible to overexpose them, but desirable background detail can be lost, with little increase in contrast, by opening the diaphragm too far. Bottom reflections can cause trouble and should always be considered. If in doubt, take several pictures at different f-stops. The added expense is small compared to the saving in effort and time. Color films have latitudes of about ½ f-stop, while black-and-white high-speed films have wider latitudes. Bracket color photos by shooting a full f-stop above and below the selected value. Go to two f-stops for bracketing black-and-white photos.

ARTIFICIAL LIGHTING

It is exhilarating to introduce artificial light underwater below the level where the reds, yellows, and oranges penetrate. Flora and fauna alike acquire a new dimension when the brilliant, dazzling colors are brought out. Artificial light is a must for the underwater photographer who goes beyond the shallow waters.

Flashbulbs have been used by underwater photographers for many years. Combined with a battery condensor unit in the camera housing or within the flash unit itself, enough current can be generated to overcome the resistance of the water.

The problem with a flashbulb system is that it is unreliable—misfirings are the rule because contacts and terminals corrode or short-out when exposed to seawater. Underwater photographers now frequently use an electronic flashing device (strobe light), which provides a quick burst of light of a much shorter duration (1/500 to 1/50,000 of a second) than the period that a camera shutter remains open. The shutter must be completely open at the time the burst of light occurs. Most 35-mm cameras have focal-plane shutters, which usually synchronize with strobe at settings of 1/60th second. Cameras with between-the-lens shutters synchronize the flash firing and shutter openings at all speeds. There are many varieties of strobes with both high and low-voltage packs. Some strobe systems have remote sensing eyes that automatically calculate the duration of light for a proper exposure.

Cameras and strobes that have proven to be good combinations in abovewater photography can be encased and used for underwater work. An experienced underwater photographer or your local "dive shop" can advise you on the best equipment to purchase.

ESTIMATING EXPOSURES USING ARTIFICIAL LIGHT

When using an artificial light source underwater, you must set up an exposure table for your particular camera/light combination, which can be used to estimate the correct f-stop setting for each situation. There are several steps that must be followed:

1. The basic abovewater formula relates distance to subject, f-stop, and guide number. The guide number is printed on the flashbulb box or the appropriate number can be selected from the back or top of the strobe unit by using the exposure dial.

Basic Abovewater Formula:

Guide number on flashbulb box or strobe unit = G
Distance to subject on land = d
f-stop setting = f

$$\frac{G}{d} = f$$

Example; for film with an ASA rating of 25:

Guide number: 80
Distance to subject on land: 10 feet

$$\frac{80}{10} = f/8$$

However, the printed guide numbers are good only on land and are at best approximate. Before you can effectively set up an underwater guide number to work with, you must determine what the *true abovewater guide number* for your own equipment is by either of two methods:

a. Take a series of exposures on land at a measured distance of 10 feet (3m) with your camera/light system, using color film, over a range of f-stop settings. Keep an accurate record. Select the photograph with the best exposure. Multiply the corresponding f-stop by 10. The product will represent the true abovewater guide number for *your* camera/light unit for the film *you* used. Remember that the guide number will vary with different ASA film ratings.

b. If a flash meter is available, place it 10 feet (3m) from the camera/light system and fire the unit. This will activate the needle on the flash meter, and it will come to rest at the correct f-stop. Note: these units do not work with flashbulbs.

2. Now to obtain your *underwater guide number*, divide the true abovewater guide number you have determined by 3.5, which is the average filter factor of water. (For extremely clear water use a value of 3; for turbid water use a value of 4.)

Underwater Guide Number Formula:

True abovewater guide number = TG
Water filter factor = WF
Underwater guide number = uwG

$$\frac{TG}{WF} = uwG$$

3. The next factor that must be determined is the *apparent underwater distance* between your camera and the subject. The apparent underwater distance is two-thirds of the true distance, so 1 apparent foot equals 16 true inches. Thus an underwater subject that appears to be 10 feet away from the camera is actually 13 feet 4 inches away. An easy way to measure apparent distance while underwater is to tie a knot every 16 inches in a line so you have a true measured distance to work with—each space between the knots is 1 apparent foot.

4. You can now calculate the correct underwater f-stop setting.

Basic Underwater Formula:

Underwater guide number = uwG
Apparent underwater distance = d
f-stop setting = f

$$\frac{uwG}{d} = f$$

Using this formula you can set up an exposure table, relating apparent distance underwater to the f-stop setting on your camera. As an example, assume you are using a strobe unit at 1/60th of a second, a film with an ASA rating of 25, having a true abovewater guide number of 90, which has been divided by the average water filter factor of 3.5 to obtain the underwater guide number of 26. Divide 26 by the apparent distance between camera and subject to obtain the f-stop setting for that distance.

Exposure table example:

UW guide number/ apparent distance in feet	*Range of f-stop settings*
26/2	f/11-f/16
26/3	f/8
26/4	f/5.6-f/8
26/5	f/4-f/5.6
26/6	f/4
26/7	f/4
26/8	f/3.5

The exposures for extremely closeup or macro-photography are normally the same as those for 1½ to 2-foot apparent distances. Some cameras employ extension tubes (macro tubes) between the lens and camera to bring a closeup subject into focus, and these reduce the amount of light that is transmitted to the film. The distance from the subject to the light is very important in macro-photography. The proper distance for your camera/light system should be determined by testing. To be absolutely certain of success, bracket the selected f-stop by also exposing shots at one f-stop on either side of the selected value.

There are two other important considerations: (1) the intensity of the light and (2) the position of the strobe or flash with respect to the subject. You can depend entirely on the artificial light source for closeup work. But as the distance from the subject increases you can begin to utilize the ambient or available light, so that the strobe/flash is used only as a fill light to bring out as much color as possible. A light meter should be used to get the camera f-stop settings, which may be the same as those calculated for the artificial light or higher if the light meter so indicates.

The second consideration, the angle at which the light strikes the subject, influences the appearance of the final picture greatly. The flash/strobe should usually not be aimed directly at a subject because any particles suspended in the water between the camera and the subject will reflect light back into the lens. Good general lighting is obtained when there is about a 45° angle between the lines from the camera and the light to the subject. Special effects can be produced by putting the light in different places. Back lighting, where the light source is behind the subject, can provide extremely beautiful results. Lighting from above tends to isolate the subject in a blaze of color. Experiment with lighting–try different angles. See what pleases you most.

MOTION PICTURES

Use a wide-angle lens of 13-mm focal length or less for ordinary underwater movies. Closeups can be done with a 25-mm lens (considered standard for topside photography). Considerably greater diaphragm openings will be needed for underwater movies than for stills. An f-stop of 2 will permit work to depths of at least 30 feet (9m) in ordinary water using film such as professional Ektachrome. A lens with a 1.0 f-stop will double this depth range or permit even deeper work under good conditions. A lens with a long depth of focus is essential for easy operation. For example, a lens with a depth of focus from 4 to 25 feet (1.2-7.5m) will usually not require further adjustment, since objects more distant than that are nearly always hazy, even in clear water.

Eight mm movie cameras are coming into extensive use by amateur divers. Operation is cheap and they are conveniently small. For projection of a large image, however, 16-mm film is most desirable, and a speed of 24 frames per second is ordinarily best. If shooting is done from an unstable situation, the speed should be increased to 32 or even 64 frames per second, light permitting.

Cameras should have a long film run per winding. It is annoying to have the motor run down in the middle of an exciting scene. Electric motors are a good, although expensive, solution to this problem. Do not descend with only a small amount of unused film left in the camera; surfacing, getting dried, and reloading use much precious time while your partner is waiting.

Avoid jerky camera movements. This will be especially difficult in shallow water, but it is absolutely essential. Brace well before starting, and if necessary weight yourself heavily; if you feel yourself tipping, stop shooting. Remember to eliminate the slight rolling movements that ordinarily arise from kicking during swimming. Those rotate the scene clockwise and then counter-clockwise when projected, and make the audience seasick.

As with still photography, maximum shooting distances should be one-fourth of the visibility distance. It is nonetheless permissible to shoot an approaching swimmer or fish beyond this range, provided the action continues until the subject is clearly in focus. When objects are moving transversely to the photographer, pan with them so they remain in the center of the field. Otherwise avoid panning because of close working distances. Zoom shots are easy; just photograph while swimming toward a stationary object. This is quickly overdone, however; include only 2 or 3 such sequences per film.

Use a fast film whenever possible, permitting smaller diaphragm openings with increased depths of focus. Hold the camera on underwater subjects a little longer than normal. Aquatic creatures may be quite unfamiliar to members of the audience, and additional time is needed for perception and recognition. Films longer than 4 or 5 minutes should include topside footage. A long series of underwater scenes becomes monotonous even to the most ardent diving enthusiasts. Cutaways to topside action here and there will greatly enhance your underwater movies.

AQUARIUM SHOTS

Many underwater photographers shun aquaria. Such work is considered unnatural, or perhaps even cheating slightly. Aquarium shots, however, contribute a great deal to understanding aquatic life. Much important work must be done in aquaria and in the studio or laboratory. Small or microscopic creatures, shy organisms, infrequent activities, animal fluorescence, time-lapse sequences, life histories, are all handled most conveniently on dry land.

Good aquarium photography is often more difficult than work in the field. All the problems of maintaining creatures in a healthy condition are combined with additional photographic difficulties. The glass wall of the tank must be clean and free from scratches. Elaborate precautions must be taken against reflections. Frequently, shots must be redone because an unsuspected ray of light has introduced flaws.

Normal behavior patterns are sometimes very difficult to elicit under aquarium conditions. It may be necessary to return to the underwater world to study what triggers a particular response and then set up an elaborate system in the aquarium to achieve these conditions. Sometimes a behavior pattern is more easily produced if it is deliberately suppressed for a while. The cleaning symbiosis, for example, may sometimes be elicited by separating the cleaning organism from the fish it cleanses. Wait until the cleaner is hungry and the fish is itching all over from its parasites. Reuniting them at this point in an aquarium should result in immediate action.

Chapter 7

COLLECTING, PRESERVING, AND IDENTIFYING SPECIMENS

COLLECTING

Mastery of collecting techniques is basic to any scientific biological work underwater. You don't have to be a dedicated biologist, however, to collect and enjoy it. Amateurs often employ the simpler techniques used by professionals. If you are a novice, start your collecting career with organisms easy to capture and to keep or preserve. Small colorful fishes are very attractive, but catching them may require lots of patience, equipment, and talent.

Man's senses are trained in a terrestrial environment, and it is necessary to learn an entirely new system of perceptual values for underwater work. Many aquatic creatures skillfully camouflage themselves; you may overlook them for a long time. Finally you become so familiar with the environment that small clues revealing the presence of a plant or animal begin to invade your consciousness. Patience and painstaking scrutiny are the earmarks of a good collector, and one often benefits by visiting a spot again and again, learning the region thoroughly and acquiring new insight as objects previously overlooked are perceived.

Once an animal is recognized, recovery may be merely a matter of grasping it, prying it off, or anesthetizing it if it moves quickly. Frequently, valuable specimens are found when the proper collecting equipment is elsewhere. It is sometimes necessary to improvise or even outthink the creature in order to obtain it. Urchin spines make fine needles, shells can be used for cutting or scraping, stones for hammering, etc.

On long dives be sure to have a large container such as a plastic bag or gunnysack so that all specimens can be accommodated. Bottles are good for delicate creatures that might be damaged if carried loose in a sack, but remember to fill bottles with water before descending, lest they implode. Nets are useful for slow-moving or dangerous creatures, and remember that many creatures burrow, necessitating digging implements. Small, quick-moving fish are most readily obtained by slurp guns (devices that suck up small creatures suddenly after releasing the trigger) or by anesthetizing with a 6% quinoline solution squirted from a rubber bulb or plastic bottle. Large fish can be speared, netted, or stunned by a bang stick (which explodes blank cartridges).

Collecting invertebrates between the high-tide mark and 1000 feet (305m) offshore from the low-tide mark requires a permit from the Department of Fish and Game (DF&G), except that a few species such as abalone, lobster, clams, etc., may be taken for sport. If you take marine animals for sport, you must have a sport-fishing license, usually obtainable at landings and at sporting-goods and diving-equipment stores. Scientific collection permits require submission of an application form, available at regional DF&G offices, that usually takes several weeks to process. If you are a student or a junior employee, your application will require the signature of a sponsor (usually a faculty member or supervisor). Both permits and licenses involve payment of a few dollars in fees.

There are also restrictions on types of gear used in collecting specimens. The restrictions by and large are designed to prevent undue damage to populations of the species–e.g., trout, salmon, Striped Bass, and Broadbill may not be taken by spearfishing, but may be taken by hook and line. Garibaldi may not be taken or possessed at all.

Numbers and sizes of species that may be taken for sport are nearly always restricted. Limits vary with season and location; regulations change from year to year. The DF&G publishes a free pamphlet, updated each year, summarizing the regulations. Reading and understanding the current regulations may save you embarrassment, inconvenience, and a fine.

Collecting is usually so much fun that it is quite easy to overdo it and gather more material than is needed or than you can possibly handle. Avoid useless collecting and over-collecting. If you have a good reason for bringing in a fine specimen, by all means do so. But if all the divers in California's waters were to remove all the plants and animals continuously because they happened to catch their fancy, undoubtedly many species would become extremely rare or disappear altogether.

PREPARATION, PRESERVATION, AND LABELING

Proper care for a collection starts at the time the items are caught, not after they have been crushed or allowed to deteriorate on the way home. Standard methods of preparation and preservation often have to be varied according to the animal, and it may be necessary to experiment with several specimens until a suitable technique evolves.

If the creature is fleshy or has considerable tissue substance, use a preservative such as alcohol or formalin. Preservatives kill bacteria and denature (modify) tissues susceptible to bacterial attack. Often it is convenient to keep the specimen permanently in the preservative.

Organisms with small amounts of tissue often can simply be dried. Where there is so much tissue that drying is unsatisfactory, denaturation with a preservative, followed by drying, may make an excellent preparation.

Sometimes it is desirable to dispose of the soft tissues, retaining only hard skeletal structures. Alkalis such as bicarbonate of soda or weak sodium hydroxide are good, especially when heated, and bleaching liquids are effective when color damage is no problem.

In some cases pleasing effects are obtained by destroying

parts of the shell or skeleton; hard uninteresting outer layers may cover beautiful mother-of-pearl of intricate design lying in the structures underneath. These can be revealed by scraping, buffing, grinding, or etching with acid. Caution! Surface structure may be all-important for identification, and even ordinary preservatives such as formalin must be avoided. Any acidic liquids, including bleaching solutions, tend to etch shells and other living structures. Color may be damaged by any of these processes.

Good permanent results can frequently be obtained by quick denaturation in alcohol, clearing briefly in Xylene, and finally mounting the specimen in plastic. When a lengthy preparation procedure is planned, select specimens with care; they should be free from imperfections such as borings, cracks, encrustations, or natural flaws.

Most seaweeds can be preserved by spreading them on mounting paper sandwiched between two sheets of blotting paper (Fig. 29). Put a strip of cheesecloth between the plant and the blotting paper to prevent sticking. Sandwich the combination between sheets of corrugated cardboard (to facilitate drying) and press for several days (see diagram). Change blotters every day or two for large, moist specimens.

Specimens should be properly labeled (unless collected solely for decoration). A label should note the collector's name, date, time, place, and depth of collection, as well as any unusual circumstances observed.

UNDERWATER IDENTIFICATION

Identifying plants and animals while underwater often involves some criteria and techniques that are different from topside procedures. The natural coloration of organisms frequently changes drastically in the blue-green illumination and bears no resemblance to standard descriptions. Finer anatomical features may be difficult or impossible to discern, and if they are essential for identification, specimens must be collected and

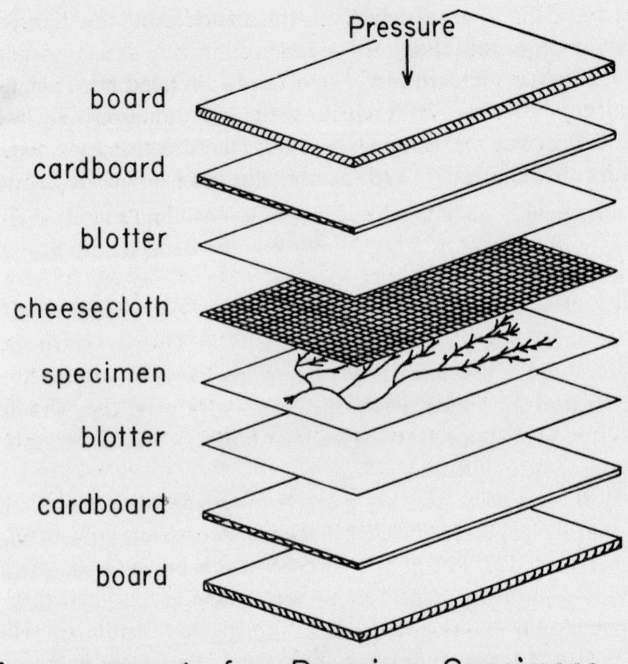

Fig. 29. Diagram showing proper arrangement of boards, blotters, and cloth for pressing and drying seaweeds on herbarium paper. The cardboard should have fluted air spaces running the entire length, to allow escape of moisture soaked up by the blotters.

examined topside—at best a tedious operation, sometimes involving the risk of losing the precise location where a species occurs. On the other hand, there are aids to identification that can be used only by divers. The preferred location of a creature is often characteristic; other examples include mode of swimming, school tendency, and growth habits. Where some unusual feature that would aid identification applies to several organisms, we have grouped them in the lists that follow (e.g., annelids with hard tubes; pelagic gastropods).

This book lists the common plants and animals of California

waters. Lack of space prevents complete descriptions. Except for small forms such as hydroids and bryozoans (often requiring use of a hand lens), most of the organisms listed can be identified underwater with practice. Individual species that are difficult to segregate while underwater are usually lumped under the genus as "several species." Emphasis has been placed on characters and groupings that the authors have found to be most useful and reliable while diving.

The diver is at a further disadvantage—he cannot take his reference book with him, and must carry any necessary information largely in his head. This is not difficult if he is seeking only one or two species. If, however, he is interested in several forms, or an entire group, or in everything he sees, a great many facts must be memorized. Often many species have several characteristics in common, and memory work can be reduced by taking this into account. For example, bony fishes have a single gill opening vs. several for the cartilaginous fishes (sharks and rays). It is thus possible to tell at a glance to which group an unknown fish belongs. It is obviously much easier to remember the group characteristic than to try to memorize the number of gill openings for each of the 22 cartilaginous fishes listed below.

The main groups can be further subdivided to reduce memory work. For example, a bony fish that is flattened horizontally in cross-section belongs to the flatfish subgroup (halibut, soles, flounders, etc.). There are only 10 common flatfishes. Hence observation of two facts (single gill opening, flattened cross-section) will greatly reduce the task of identifying a flatfish. Instead of recalling all information about 114 species, you now need remember information about only 10 species. But even 10 species requires quite a bit of memory work. It is often easiest to be familiar with the principal characteristics used for identification within a subgroup (body shape, coloration, etc., for flatfishes), note these features on the specimen you have found, and later determine the species when you can study the reference book.

Proper identification begins, of course, with the largest

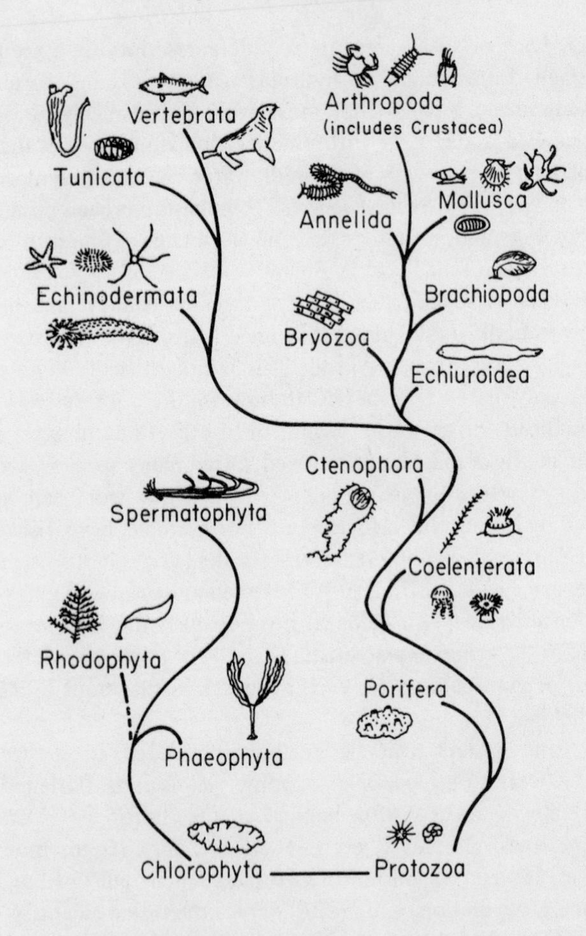

Fig. 30. Diagram showing the presumed ancestral relationships among the major phyla found in the sea.

categories of identification and proceeds to the lesser categories. First decide whether a specimen belongs in the plant or animal kingdom. For beginners this is not as easy as it may sound. In the sea there are many sessile (remaining in one location) animals that resemble plants. After the proper *kingdom* has been determined, the next-lower category is the *phylum*. For the larger phyla, selection of the correct *class* within the phylum is necessary, and for one very large class (the crustaceans), the proper *order* must be chosen. From these categories one can proceed to the *subgroups* listed in detail in the descriptions that follow.

This may strike the beginner as a rather involved procedure, not particularly appealing, and certainly not necessary if his interest is chiefly in fishes, seashells, red algae, etc. Many of the better known groups of animals such as snails, bivalves, starfish, and others are familiar, easy to recognize, and require no complex decisions for correct assignment. Indeed, the beginner will probably do well to start within a group he finds interesting and learn the species. If his interests widen, however, and he wants to know more of the fauna and flora, the higher categories of classification must be learned. After these categories and their characteristics are known, assignment of an unknown specimen to its correct group is usually done rapidly, generally much more easily than identification to species. Figure 30 shows the principal major groups in California waters.

In the next five chapters we have listed a number of the commonest genera and species found subtidally along the California coast. In numerous instances we have cited outdated latin names in parentheses as well as the current nomenclature. This should assist the reader in finding various genera and species in some of the older reference works we have cited. Many of these old volumes are quite useful for illustrations and descriptions. They should not be ignored.

Chapter 8

MARINE PLANTS

DIVISION CHLOROPHYTA–THE GREEN ALGAE

The green algae range from microscopic suspended single cells to large flat sheets of tissue. We are concerned here only with the attached forms, sufficiently large to be seen easily with the naked eye. Almost without exception, Chlorophyta are green. There are, however, marine plants which are green but belong to other groups (Rhodophyta, or red algae; also sea grasses), so that coloration is not an infallible guide to classification.

Green algae abound in bays and often are dominant in sloughs and estuaries. They are usually present in coves and sheltered waters of the open coast, but only a few species of Chlorophyta can withstand violent wave action in exposed situations. Most species of the group seem to prefer high light intensities, and inhabit shallow waters. Consequently they do not become abundant in deep water where wave surge is reduced. One genus, *Enteromorpha*, often flourishes where human wastes are dispersed into natural waters–it has thus been used as an indicator of pollution, but it also often occurs in pristine waters.

Bryopsis corticulans: dark green, up to 3 inches (7.6cm) tall; resembles a tiny pine tree. Occasional, entire coast, shallow.

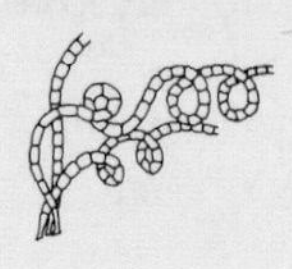

Chaetomorpha torta: blue-green; appears as a stiff, segmented, long tube entwined around other plants or objects. Southern California and Baja California, intertidal to 70 feet (21.5m).

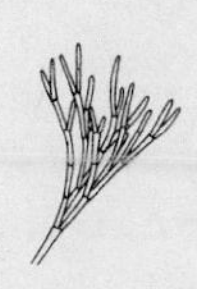

Cladophora graminea (and other species): forms bright green tufts up to 3 inches (7.6cm) tall, especially where rock and sand overlap. Central California to Baja California, intertidal to 70 feet (21.5m).

Codium fragile: dark green; spongy, evenly forking, cylindrical branches; up to 1 foot (30.5cm) tall. *C. cuneatum*: similar, but slightly flattened in cross-section. Entire coast, shallow.

Codium setchellii (and *C. hubbsii*): dark green; irregular, spongy nodules, lobes, or pancake-shaped growths. Entire coast, intertidal to 50 feet (15m) or more.

Enteromorpha (several species): bright green, unbranched or branching tubes or cylinders, usually collapsed and resembling long delicate ribbons. Common in bays and quiet waters; entire coast, shallow.

Halicystis ovalis: bright green sphere, ½-inch diameter or less; appears black at substantial depths. Entire coast, intertidal to 75 feet (23m).

Monostroma (several species): yellow- to brownish-green, membranous, extremely delicate sheets of tissue; usual span 1 or 2 inches (2.5 or 5.1cm) but occasionally much larger. Oregon border to Monterey, shallow.

Ulva (Sea Lettuce; several species): green, membranous, up to 1 foot (30.5cm) tall in sea, larger in bays. Entire coast, intertidal to more than 30 feet (9m).

DIVISION PHAEOPHYTA–THE BROWN ALGAE

Brown algae dominate much of the exposed California coastline, either as large treelike plants or as shrublike forms. A few Rhodophyta are also brown; hence coloration as an aid to classification must be used with some restraint. Even so, if a seaweed is brown the chances are very good it belongs to the Phaeophyta.

The subdivisions we have established are useful for rapid characterization in the field, but do not necessarily reflect close relationships in a genetic sense. The reader who wants to know more about evolutionary ties within the brown algae will need to consult the standard texts on seaweeds listed in our Bibliography.

MAJOR SUBDIVISIONS BY GROSS APPEARANCE

I. Large treelike plants, usually bearing gas bulbs for flotation
II. Medium height, with straplike blades crowning a conspicuous *stipe* (stalk) 2 to 4 feet (0.75-1.5m) high
III. Low height, but long blades sprawling horizontally
IV. Small plants, various shapes

I. LARGE TREELIKE PLANTS

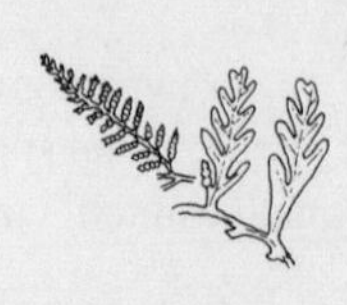

Cystoseira osmundacea (reproductive phase, spring and summer): stringy yellow network with small pods, arising from shrublike brown plant. (Nonfertile phase described in group IV below.) Entire coast, intertidal to 100 feet (30.5m) or more.

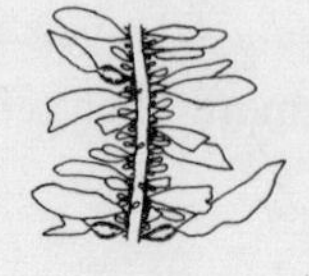

Egregia (Feather-Boa Kelp): long straplike stipes bordered with bladelets, giving ruffled or fuzzy appearance; pear-shaped bulbs. *E. menziesii* north of Point Conception, *E. laevigata* (illustrated) from Point Conception to Ensenada, *E. planifolia* south of Ensenada; intertidal to 50 feet (15m).

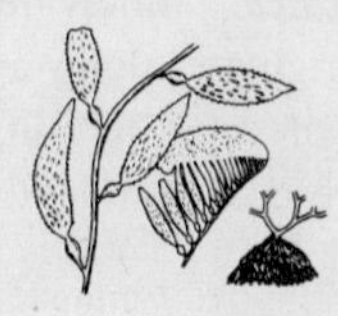

Macrocystis (Giant Kelp): pointed corrugated blades attached by pear-shaped bulbs to vinelike stipes up to 200 feet (61m) long. *M. integrifolia* from Kibesillah Beach to San Simeon; *M. angustifolia* from Monterey to San Clemente; *M. pyrifera* (illustrated) south of Oceanside; intertidal to 130 feet (39.5m) (see also Plate 1b).

Nereocystis leutkeana (Bull Kelp): streamerlike blades fastened to baseball-sized float which tapers to ropelike stipe; up to 100 feet (30.5m) tall. Oregon border to San Miguel Island, 10 to 100 feet (3-30.5m).

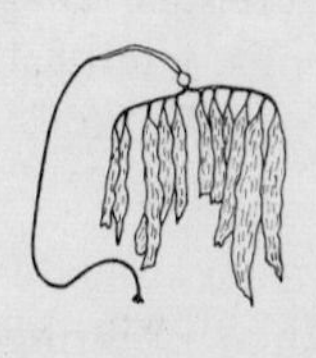

Pelagophycus (Elk Kelp): antler-shaped branches join huge corrugated blades to melon-sized float; tapering ropelike stipe; up to 80 feet (24.5m) tall. *P. porra* (illustrated) off southern California and Baja California mainland; *P. giganteus* off Catalina and San Clemente islands; 10 to 90 feet (3-27.5m) or more.

II. MEDIUM HEIGHT STRAPLIKE BLADES, CONSPICUOUS STIPE

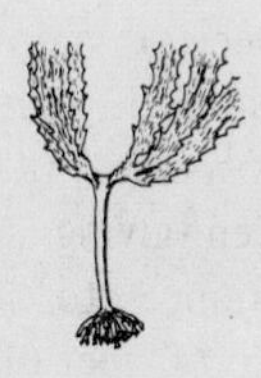

Eisenia arborea: stipe forks below origin of corrugated blades. Monterey to Baja California, intertidal to 120 feet (36.5m).

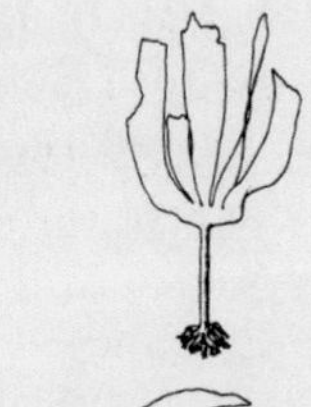

Laminaria setchellii: undivided stipe; smooth blades originate at apex. Entire coast, intertidal to 30 feet (9m).

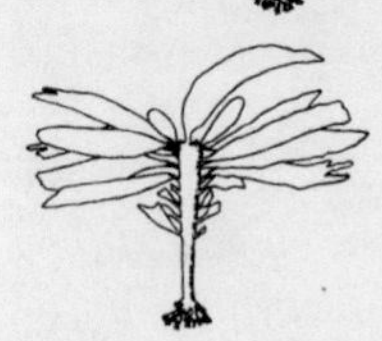

Pterygophora californica: undivided stipe; smooth blades originate along approximately upper half of stipe, terminal blade has midrib. Entire coast, intertidal to 120 feet (36.5m).

III. LOW, WITH LONG SPRAWLING BLADES

Agarum fimbriatum: dark brown, corrugated blade, conspicuous midrib, often pocked with holes; bladelets fringe edges of short stipe. Entire coast, 40 to 100 feet (12-30.5m).

Alaria marginata: tan, smooth surface, lacerated edges on main blade with rounded midrib; small rounded blades emerge laterally from short stipe. Oregon border to central California, intertidal to shallow subtidal.

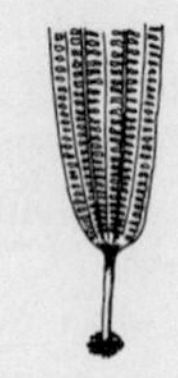

Costaria costata: dark brown, single corrugated blade with five midribs; up to 6 feet (2m) long. Oregon border to southern California, shallow.

Desmarestia (several species): yellowish-green to dark brown; smooth, complex branching (except *D. tabacoides*, below), tapering branchlets often giving superficial segmented appearance to plant. Entire coast, intertidal to 60 feet (18m). *D. munda* illustrated.

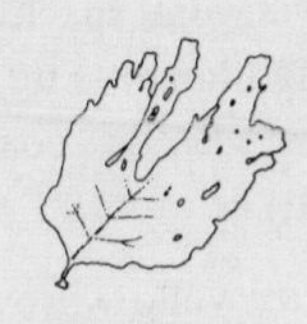

Desmarestia tabacoides: dark greenish-brown, smooth, usually single blade, but often torn to form secondary blades; faint veins, short stipe; outline roughly circular. Central California to Baja California, 40 to 135 feet (12-41m).

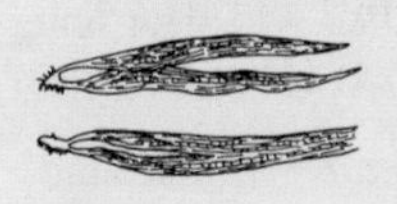

Dictyoneurum californicum: yellow to brown, light ribbing subdivides blade into rectangular areas; up to 3 feet (1m) long. Oregon border to central California, intertidal and shallow subtidal.

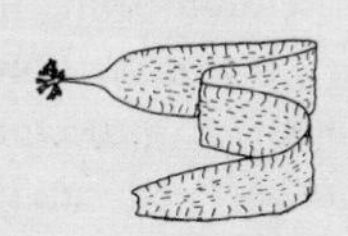

Laminaria farlowii: dark brown, corrugated blade; up to 14 feet (4.5m) long. Central California to Baja California, intertidal to 100 feet (30.5m).

IV. SMALL PLANTS, VARIOUS SHAPES

Colpomenia sinuosa: yellow-green; irregular, roundish, bulbous or convoluted masses. Entire coast, intertidal to 50 or 60 feet (15-18m).

Cystoseira osmundacea (nonreproductive phase): dark brown, tough woody stem. Four to 12 inches (10.2-30.5cm) high. (See group I for reproductive phase.)

Dictyopteris zonarioides: yellow-brown bushes, sometimes iridescent blue, up to 1 foot (30.5cm) high; blades with midrib. Southern California and Baja California, intertidal to more than 100 feet (30.5m).

Dictyota flabellata: yellow clumps up to 5 inches (12.7cm) high; small holes pierce lobed blades along edges. Central California to Baja California, intertidal to 80 feet (24.5m).

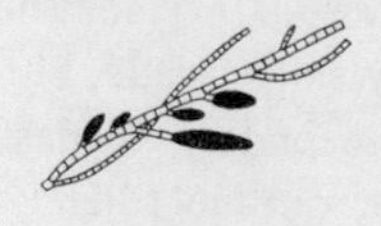

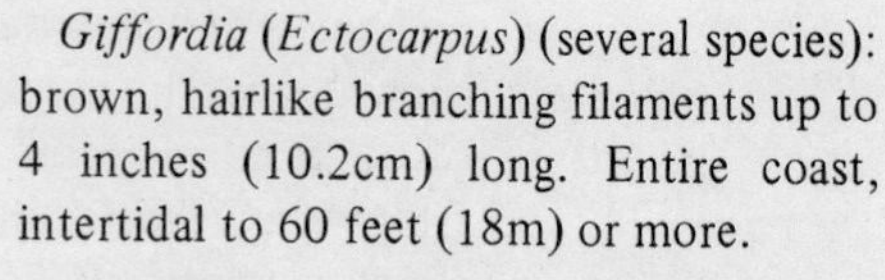

Giffordia (*Ectocarpus*) (several species): brown, hairlike branching filaments up to 4 inches (10.2cm) long. Entire coast, intertidal to 60 feet (18m) or more.

Pachydictyon coriaceum: yellow-brown clumps up to 1 foot (30.5cm) high; tiny holes randomly across bases of lobed blades. Southern California and Baja California, intertidal to 60 feet (18m).

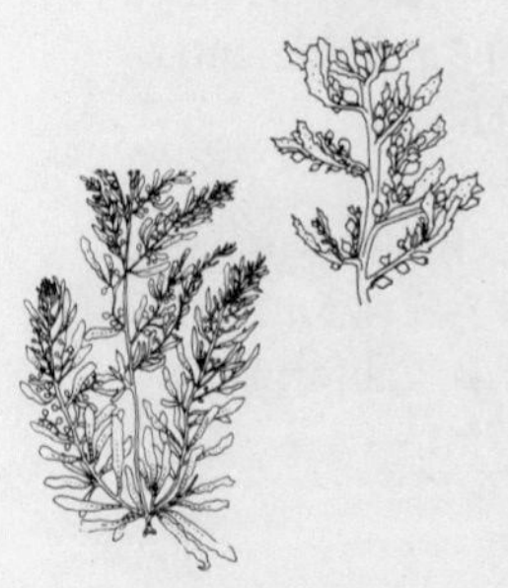

Sargassum agardhianum (illustrated, top): abundant serrated (holly-like), undivided bladelets interspersed with small spherical floats. Southern California to Baja California, shallow. *S. palmeri*: nonserrated, divided bladelets. *S. muticum* (illustrated, bottom): nonserrated, undivided bladelets. Entire coast in quiet water.

Taonia lennebackerae: yellow, thin, lacerated blades with striations near tip, tapering to narrow base; up to 2 feet (61cm) high. Southern California and Baja California, intertidal to 40 feet (12m).

Zonaria farlowii: dark bushy clumps up to 1 foot (30.5cm) high; fan-shaped blades with light margins and concentric striations. Southern California and Baja California, intertidal to 100 feet (30.5m).

DIVISION RHODOPHYTA–THE RED ALGAE

Red algae are usually the most difficult to identify. The color is not necessarily red. Green, yellow, purple, brown, black, and pink, as well as all shades of red, characterize the group. Fortunately, size is a help.

None of the reds are as big as the large kelps, none are corrugated, and only two larger common subtidal reds (both *Iridaea*, sometimes reaching lengths of intermediate-sized kelps) might be mistaken for brown or green algae because of their color. Hence a seaweed that is brown, green, or yellow and over 12 to 18 inches (30.5-45.7cm) long, probably belongs to the Chlorophyta or Phaeophyta. Among the reds smaller than 12 inches (30.5cm), *Gigartina canaliculata* is most frequently mistaken by amateurs for a brown or green algae. If the novice, therefore, will take the trouble early in his career to become familiar with *G. canaliculata, Iridaea flaccida*, and *I. lineare*, he can be fairly sure that most of the other Rhodophyta he sees below the intertidal will either be black or have a sufficient tinge of red to be recognizable as red algae.

Rhodophyta are so diverse that even advanced amateurs must often be content with identification only to the genus level. Some of the easily recognizable species are listed below, but in many instances only general generic characters are listed. Again, our subdivisions are useful for field classification but do not imply close evolutionary relationships.

MAJOR SUBDIVISIONS BY GROSS APPEARANCE

I. Foliose forms–plants with thin, flat, soft blades; blade width is many times the thickness
II. Arborescent forms–plants mainly composed of cylindrical or slightly flattened stalks or branches, or of hard calcified segments
III. Filamentous forms–plants entirely composed of fine filaments
IV. Crustose forms–plants forming flat, ridged, or lumpy encrustations on rock, shell, or other plants

I. FOLIOSE FORMS–THIN, FLAT, SOFT BLADES

A. Blades round, oval, or pear-shaped; upper border rounded or a blunt point
B. Adult blade elongate, length two or more times width, usually tapering at both ends to a point

C. Blades form a branching system of fingerlike projections or lobes; lateral margins of branches smooth
D. Blades form a branching system of fingerlike projections, lobes, or pointed blades; lateral margins of branches not smooth (ruffled, cusped, proliferous bladelets, etc.)
E. Irregularly shaped blades

A. Blades Round, Oval, or Pear-shaped

Cryptonemia (several species): red, smooth, sometimes membranous blades, easily confused with *Porphyra* and *Halymenia* (group IB, below). *C. obovata* (illustrated) fairly common, often multi-lobed; 1 foot (30.5cm) or more high. Entire coast, 10 to 100 feet (3-30.5m).

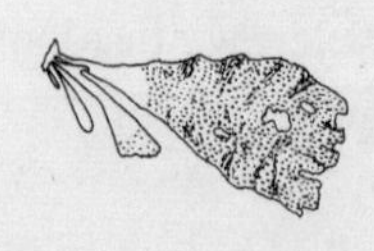

Gigartina (several species): various shades of red; thick, tough blades, easily distinguished by papillae on blade surface. *G. corymbifera* common north of Point Conception, *G. exasperata* (illustrated) along entire coast, intertidal to 30 feet (9m). (For other species, see groups IB and IIA below.)

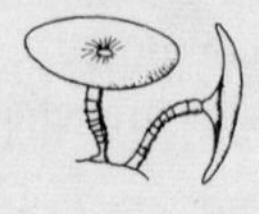

Maripelta rotata: purple-red, iridescent, delicate dollar-sized blade, centrally stalked like a lily-pad. Monterey to Baja California, 50 to 200 feet (15-61m).

Opuntiella californica: red-to-black, formed of tough, pancake-sized blades joined at edges. Entire coast, intertidal to 80 feet (24.5m).

Smithora naiadum: reddish-purple, transparent bladelets growing on eel grass (*Zostera*) or surf grass (*Phyllospadix*) in cold-water areas. Entire coast, shallow, spring and summer.

B. Adult Blade Elongate

Erythrophyllum delesserioides: red blade, frequently torn from edge to midrib; conspicuous midrib. Oregon border to central California, intertidal to about 20 feet (6m).

Gigartina (several species): red, purple, or black, tough blades up to 3 feet (1m) long; surface typically papillated, sometimes proliferous bladelets at edges. Entire coast, intertidal to 30 or 40 feet (9 or 12m). *G. agardhi* illustrated.

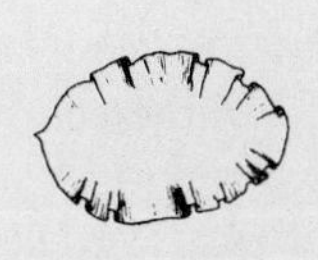

Halymenia: cherry red; smooth, membranous, 2 feet (61cm) long. *H. coccinea* (illustrated) fairly common; surface slippery, distinguishes from *Porphyra* (below) and *Cryptonemia* (group IA, above). Entire coast, intertidal to 120 feet (36.5m).

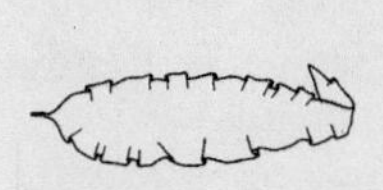

Iridaea (several species): red or purple except for brown *I. lineare* and green *I. flaccida*; lanceolate except for multilobed *I. heterocarpum*. Toughness distinguishes from *Porphyra* (below), *Cryptonemia* (group IA, above), and *Halymenia* (above). Entire coast, intertidal and shallow. *I. splendens* illustrated.

Porphyra (several species): variously red to purple (wider color range intertidally); delicate, membranous, up to several feet long. Entire coast, intertidal to 100 feet (30.5m). *P. perforata* illustrated.

C. Blades Form a Branching System with Smooth Lateral Margins

Callophyllis (several species): red, smooth, fairly delicate; branchlet tips typically dissected into points, cusps, or tiny lobes. Entire coast, intertidal to 80 feet (24.5m). *C. flabellulata* illustrated.

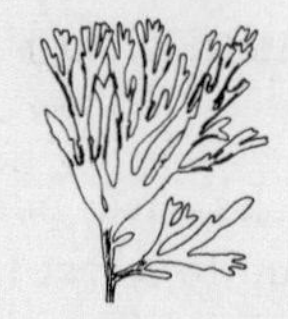

Cryptopleura violacea: purple-olive to red, smoothish, delicate; branchlet tips rounded, with veins extending nearly to tips (hold up to light to see). Entire coast, intertidal to 30 feet (9m).

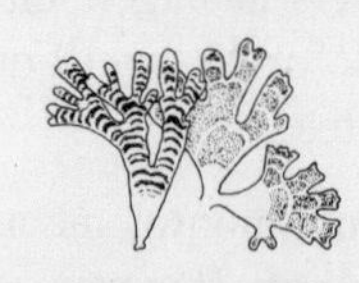

Fryeella gardneri: red, sometimes iridescent blue, delicate, smooth; tips blunt or rounded, sori form "thumbprint pattern" on branch surfaces (hold up to the light). Entire coast, 20 to 150 feet (6-45.5m).

Polyopes bushiae: deep red, up to 1 foot (30.5cm) high; branching tips about ¼ inch (0.6cm) across, slightly recurved; resembles a small bush. Occurs south of Newport Bay in warmer sites (inlets, etc.) from 15 to 50 feet (4.5-15m) deep.

Rhodoglossum affine: dark red, smooth, blades with round, obtuse, or acute tips concavely curved; proliferous bladelets fringing basal segments. Central and southern California, shallow.

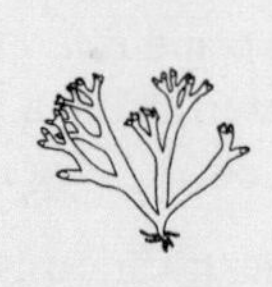

Rhodymenia (several species): red, delicate, broadly rounded branchlets, tapering basally to round stipe; branching pattern often yields fan-shaped plant. Entire coast, intertidal to 100 feet (30.5m). *R. californica* illustrated.

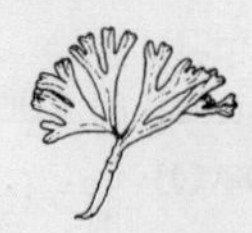

Stenogramme interrupta: red, smooth, rounded tips, distinguished from *Rhodymenia* (above) by midrib-like line of sori or irregular blotching (on fertile plants only). Entire coast, intertidal to 130-135 feet (40-41m).

D. Blades Form a Branching System, Lateral Margins Not Smooth

Acrosorium uncinatum: bright red, smooth, commonly growing entangled in other plants; branch margins often hook-shaped. Central California to Baja California, intertidal to 125 feet (38.5m).

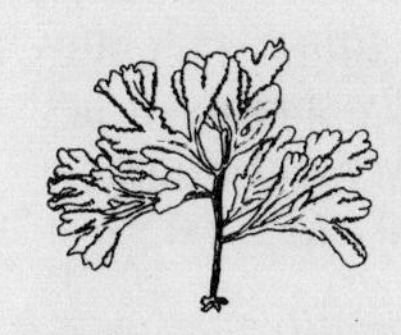

Botryoglossum farlowianum: red to purple, iridescent; ruffled margins, usually light veins, up to 16 inches (40.6cm) high. Often dominates shallow bottom in cold areas. Entire coast, intertidal to 40 feet (12m).

Cryptopleura lobulifera: margins lobed. *C. crispa*: ruffled margins, otherwise closely resembling *C. violacea* (see group IC, above). Distinguished from *Botryoglossum* (above) by brown-to-olive shading in color. Entire coast, intertidal to 60 feet (18m).

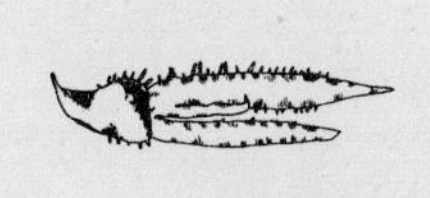

Grateloupia (several species): delicate, red, usually veined surfaces, margins ruffled or lobed; generally small plants growing at or around bases of kelps. Northern and central California, shallow. *G. californica* illustrated.

Nienburgia andersoniana: red, thin, elongate branches with points or cusps along margins. Central California to Baja California, intertidal to 60 feet (18m).

Prionitis (several species): red-brown to almost black, tough, narrow, pointed blades with numerous pointed bladelets along edges. Entire coast, intertidal to 30 to 40 feet (9-12m). *P. linearis* illustrated.

E. Irregularly Shaped Blades

Polyneura latissima: pink to red, fairly delicate; conspicuous veins, blade tapering at base. Entire coast, intertidal to 60 feet (18m).

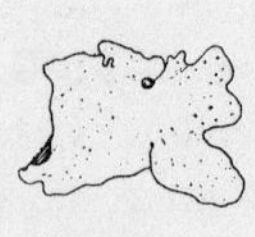

Pugetia (*Callophyllis*) *firma*: red, slippery, tough, blade usually sprawls or lies close to bottom; roughly circular outline. Entire coast, intertidal to 100 feet (30.5m).

Schizymenia pacifica: red-brown, slippery to slimy, fairly tough; outline roughly oval. Entire coast, intertidal to 50 feet (15m).

Weeksia reticulata: pink to red, fairly delicate; oval to fan-shaped, tapering at base, reticulate network of veins (hold up to light to see). Cold areas, entire coast, 10-20 feet to 100 feet (3-6m-30.5m).

II. ARBORESCENT FORMS–CYLINDRICAL STALKS, OR CALCIFIED SEGMENTS

A. Ramified branching produces complex tangle in upper portions; soft tissues

B. Upper portions not tangled, or are easily separated; soft tissues

C. Entire plant formed of small, hard, calcified segments

A. Complex Tangled Branching; Soft Tissues

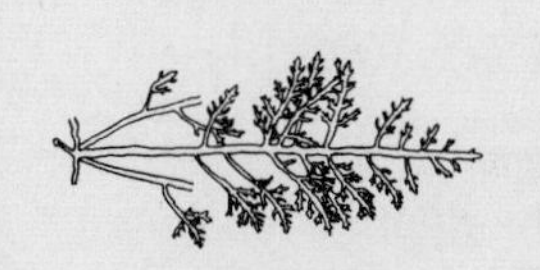

Gelidium (several species): red to purple, up to 2 feet (61cm) high. Smallest branchlets often bend acutely near attachment point, otherwise closely resembles *Microcladia* and *Pterocladia* (below). Entire coast, intertidal to more than 60 feet (18m). *G. robustum* illustrated.

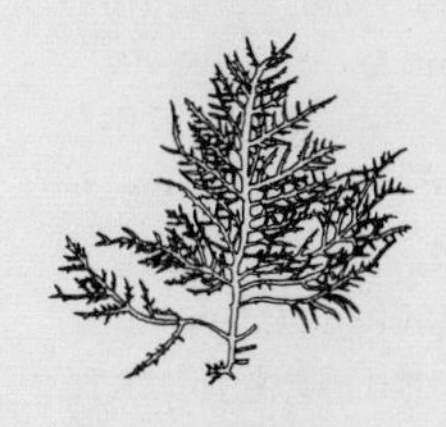

Gigartina canaliculata: yellow to dark green, up to 1 foot (30.5cm) high; smallest branchlets give thorny appearance. Often dominant at mean tide level, rapidly thinning subtidally; entire coast, to 30 feet (9m).

Laurencia (several species): green-or-purplish red to red, up to 1 foot (30.5cm) tall; branchlet tips often appear stubby, sometimes pitted. Entire coast, intertidal to 50 feet (15m). *L. spectabilis* illustrated.

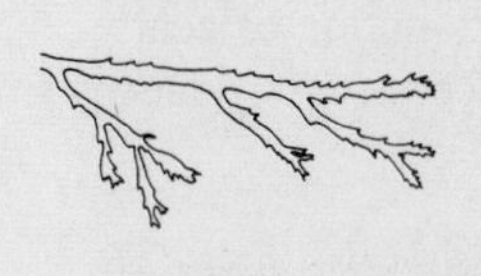

Leptocladia: dull red, up to 1 foot (30.5cm) tall; shoots and branches flat at edges in cross-section, bulge in center. *L. conferta* northern California, *L. binghamiae* (illustrated) southern California; intertidal to 60 feet (18m).

Microcladia coulteri: deep red, 1 foot (30.5cm) or more high, often attached to other plants; foliage on branches tapers evenly towards tips, like a pine tree. Entire coast, intertidal to 30 feet (9m).

Plocamium coccineum: bright red, up to 1 foot (30.5cm) high; foliage on branches does not taper evenly toward tip. Recurved branchlets resemble crab pincers at branch tips. Entire coast, intertidal to 80 feet (24.5m).

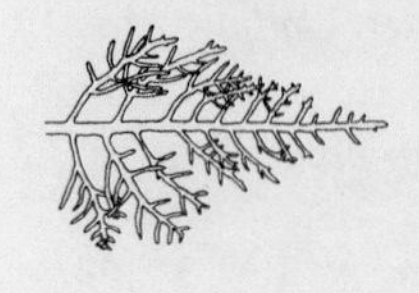

Pterocladia pyramidale: dark red, up to 1 foot (30.5cm) high; elliptical or flattened in cross-section (unlike *Microcladia* and most *Gelidium* species–see above). Southern California and Baja California, intertidal to 50 feet (15m).

Ptilota and *Neoptilota* (several species): various shades of red, up to 1 foot (30.5cm) tall; main branches flattened or compressed, fringed with tiny leaflets giving fuzzy appearance to edges. Entire coast, shallow. *P. californica* illustrated.

B. Simple Branching; Soft Tissues

Agardhiella tenera: bright red, often 1 foot (30.5cm) or more high; branches lumpy or smooth, tapering evenly to a point. Often occurs at junction of coarse sand and rock; entire coast, intertidal to 80 feet (24.5m).

Gracilaria verrucosa: purple with yellowish tips, up to 16 inches (40.7cm) high. Occasional, in quiet water at junction of coarse sand and rock; entire coast, intertidal to shallow subtidal.

Gracilariopsis sjoestedtii: yellow to reddish purple, up to 6 feet (2m) tall, but usually less than 1 foot (30.5cm); cylindrical pointed branches, often bumpy due to reproductive cystocarps. Entire coast, intertidal to 60 feet (18m) or more.

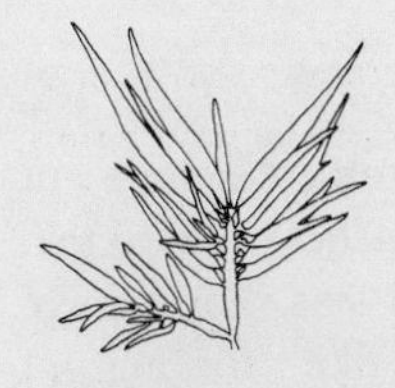

Prionitis (several species): dark red, tough shoots, 1 foot (30.5cm) or more tall; branches often fork just below tips. Entire coast, intertidal to 60 feet (18m). *P. andersoniana* illustrated.

C. Plant Formed of Small Calcified Segments

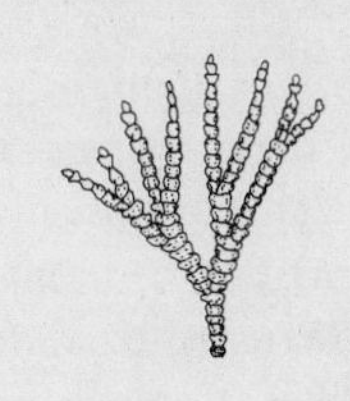

Bossiella (several species): reddish-pink, up to about 6 inches (15.2cm) tall; arrowhead-shaped segments, with small bumps (reproductive conceptacles) near center. Entire coast, intertidal to 130 feet (40m).

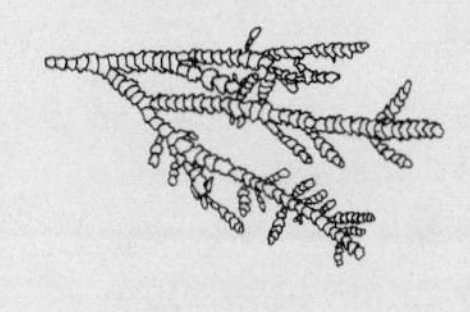

Calliarthron (several species): reddish-purple, up to 1 foot (30.5cm) tall; flattened or round segments, with small bumps (reproductive conceptacles) centrally and on edges. Branching usually simpler than *Bossiella* (above) or *Corallina* (below). Common where sea urchins dominate; entire coast, intertidal to 70 feet (21.5m). *C. cheilosporioides* illustrated. (See also Plate 1e.)

Corallina (several species): pink-to-reddish purple, height up to 6 inches (15.2cm); segments cylinders or truncated cones. Proliferous branchlets give featherlike appearance. Entire coast, intertidal to 70 feet (21.5m). *C. officinalis* var. *chilensis* illustrated.

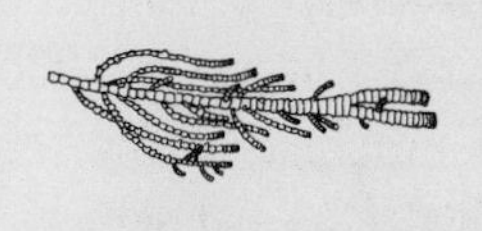

Lithothrix aspergillum: pinkish-gray, up to 5 inches (12.7cm) tall; cylindrical segments broadening toward tip, which is blunt. Often found where sand intrudes into boulder patches; entire coast, intertidal to 40 feet (12m).

III. FILAMENTOUS FORMS

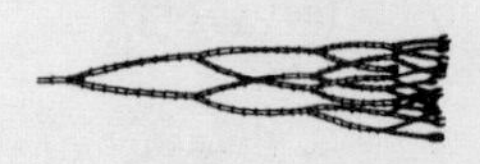

Centroceras clavulatum: blackish- or brownish-red, up to 3 inches (7.6cm) tall. Examination with hand lens shows cross-banding and pincers-like forking at branchlet tips. Often growing on other plants; central California to Baja California, intertidal to shallow subtidal.

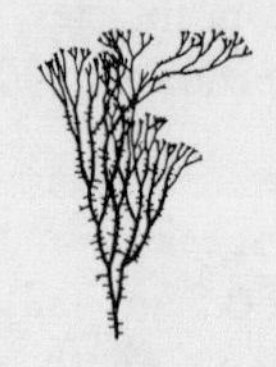

Ceramium pacificum: deep red, occasionally to 6 inches (15.2cm) but usually less than 2 inches (5.1cm) tall. Examination with hand lens shows cross-banding and forcipate tips like *Centroceras* (above), but *C. pacificum* has many short lateral branchlets in lower portions. Often growing on other plants; entire coast, intertidal to 30 feet (9m).

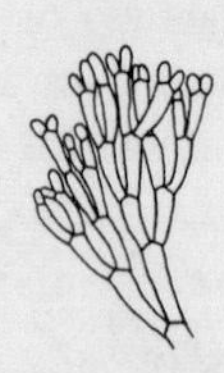

Griffithsia pacifica: reddish-pink tufts, up to 4 inches (10.2cm) tall; no forcipate tips as in *Centroceras* and *Ceramium* (above). Branchlets formed of single cells (use hand lens). Occasional; entire coast, intertidal to 100 feet (30.5m).

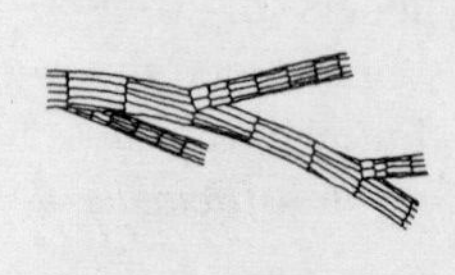

Polysiphonia (several species): brown-to-blackish red, usually less than 2 inches (5.1cm). Branchlets pointed and formed of many cells (use hand lens). Grows on other plants; entire coast, intertidal to 50 feet (15m). *P. californica* illustrated.

IV. CRUSTOSE FORMS–ENCRUSTATIONS ON ROCK, ETC.

Dermatolithon (several species): chalky pink; forms round encrustations up to ½-inch (1.3cm) in diameter on flat surfaces of host plants, or may encircle cylindrical branches. Entire coast, intertidal to 30 feet (9m). *D. dispar* illustrated.

Lithophyllum (several species): pink or whitish-pink forms; smooth to slightly bumpy round encrustations on rock and shell. Entire coast, intertidal to 75 feet (23m). *L. imitans* illustrated.

Lithothamnium (several species): whitish-pink to gray to purplish-red, smooth to highly convoluted encrustations or overlapping ridges on rock and shell. Entire coast, intertidal to more than 70 feet (21.5m). *L. lamellatum* illustrated.

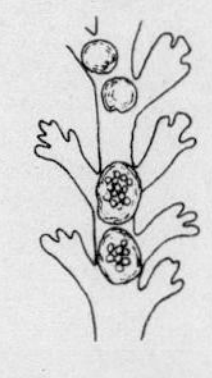

Melobesia marginata: purple; small button-shaped encrustations on plants. Entire coast, shallow.

Melobesia mediocris: pink; small encrustations on surf grass (*Phyllospadix*) blades. Entire coast, intertidal to about 30 feet (9m).

Peyssonelia (several species): bright to dark red, forms soft or hard, roughly circular encrustations on rock and shell. Entire coast, intertidal to 60 feet (18m) (Plate 8b).

PHYLUM SPERMATOPHYTA–FLOWERING PLANTS

A few higher plant species with flowers, seeds, and roots occur on the California coast and in embayments. Frequently they dominate their territories, to the virtual exclusion of the algae.

Phyllospadix (Surf Grass): bright green, ribbonlike, tough blades. *P. scouleri*, in surf zone, has thinner, broader blades than *P. torreyi* (illustrated), which extends down to 40 feet (12m) occasionally. On rocks; entire coast. (See also Plate 8c.)

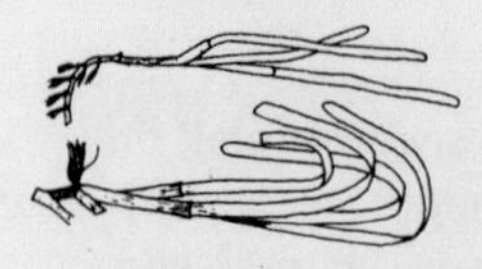

Zostera marina (Eelgrass): green, ribbonlike blades, about ¼-inch (0.7cm) width. Common in bays to 20 feet (6.1m), occasional on sand bottoms of open coast areas to 40 feet (12m); entire coast.

Chapter 9

MARINE ANIMALS: PROTOZOA THROUGH BRACHIOPODA

PHYLUM PROTOZOA

Protozoans are the most primitive animals. Within the structure of a single cell they combine the essential functions required for living. Some protozoans appear as cellular aggregates, but each component cell could survive independently if necessary. Most are microscopic, but a few are sufficiently large to be seen with the naked eye or with a hand lens. Some of the larger protozoans belonging to the order Foraminifera have shells (*tests*), are sessile, and are first perceived attached to stalks, stipes, etc., when you are combing through the day's collection at home. One "foram," *Gromia*, is sufficiently large to be perceived easily in the field, through a face mask.

Discorbis: chambered test is spiral, resembling a tiny snail. Often attached to hydroid stalks. Probably entire coast, intertidal to 60 feet (18m).

Folliculina: often blue; test resembles a vase or bottle. Entire coast, intertidal to 60 feet (18m).

Gromia oviformis: tan to brown, shiny small sphere up to 1/8-inch (0.3cm) in diameter. Entire coast, intertidal to 100 feet (30.5cm).

PHYLUM PORIFERA–THE SPONGES

Simplest of the multicellular animals, adult sponges are sessile, occurring as thin sheets, irregular masses, or defined shapes attached to rock, shell, wood, and other solid substrate. Even experienced collectors sometimes have trouble distinguishing between sponges and colonial tunicates in the field. Beginners may be bothered additionally by resemblances to thick bryozoan and plant encrustations. Fortunately many sponges are easily recognized, and there are reliable guidelines to help distinguish the group from other lumps and blobs that encrust the ocean floor:

1. Sponges frequently have one or more apertures (*oscules*) used for sea-water exchange. In undisturbed animals, the oscules of many sponges are clearly visible to a diver at close range. The natural openings of bryozoans, many colonial tunicates, and crustose plants are microscopic.

2. Most sponges are flexible. Even thin sheets of encrusting sponges yield slightly to finger pressure (the term "spongy" derives from this characteristic). Nearly all bryozoans and crustose plants are hard and brittle (notable exceptions: the bryozoan *Flustrella* and the flat forms of the green alga *Codium*).

3. The apertures of sponges usually close slowly (ten seconds to a minute) when the animal is handled roughly. The apertures of solitary tunicates usually close rapidly. Use this rule cautiously; very occasionally there are sluggish tunicates and easily irritated sponges that overlap in this characteristic.

4. When torn or cut to expose a cross-section of the body, the texture of most sponges appears almost uniform. Color sometimes changes from exterior to interior. You will find soft tissue interspersed with fibers or a network of tiny needles (*spicules*), but the mixture occurs throughout. Colonial tunicates in cross-section show a variety of discrete tissues and organs, not a uniform mixture.

MAJOR SUBDIVISIONS BY GROSS APPEARANCE

I. Flattened encrustations, thin to thick

II. Tubelike, ovoid, or vase-shaped individuals or groups

III. Animal approximately spherical

IV. Conspicuous upright lobes, or fingerlike or leaflike projections

I. FLATTENED ENCRUSTATIONS

Acarnus erithacus: scarlet, firm, up to 1 inch (2.5cm) thick, several inches (cm) in diameter; craterlike oscules. Entire coast, intertidal to 100 feet (30.5m) (Plate 1a).

Astylinifer arndti: yellow, orange, and purplish tinges; soft, forms thin sheets 1/16th-inch (0.15cm) thick on rocks and scallop shells. Central California to Baja California, intertidal to more than 100 feet (30.5m).

Axinella mexicana: bright red, firm, slippery-smooth crusts, up to 1 inch (2.5cm) thick; oscules (often closed) up to ¼-inch (0.6cm) in diameter. Yellow-orange interior. Southern California and Baja California, 20 to 50 feet (6-15m).

Cliona celata: yellow; bores 1/8-inch (0.3cm) holes in abalones, barnacles, etc., spreading over the surface also. Entire coast, intertidal to 130 feet (40m).

Cyamon argon: red (orange at depth), firm, wrinkled encrustations, up to 1 inch (2.5cm) thick. Exudes mucus when collected. Central California to Baja California, 20 to 50 feet (6-15m).

Esperiopsis originalis: reddish-brown, stiff, brittle, thin small patches to biscuit-sized cakes; round, unraised oscules. Entire coast, intertidal to 40 feet (12m).

Geodia mesotriaena (older specimens): dirty white surface, drab yellow subsurface; moderately soft, roundish cakes up to 3 inches (7.6cm) thick, 8 inches (20.3cm) across. Entire coast; a deep-water form, 90 feet (27m) or deeper.

Halichondria panicea: dull orange or yellow, soft to firm, encrustations up to 1 inch (2.5cm) thick; structure porous like bread. Entire coast, intertidal to at least 50 feet (15m).

Haliclona lunisimilis: yellow-tan, soft sheets ¼-inch (0.6cm) thick, with tubular oscules projecting another ¼-inch (0.6cm). *H. permollis* similar except often lavender, sometimes with no oscules. Central California to Baja California, intertidal to 40 feet (12m).

Hymenamphiastra cyanocrypta: cobalt blue, moderately soft, thin encrusting sheets; oscules obscure. Central California to Baja California, intertidal to 100 feet (30.5m).

Hymeniacidon sinapium: orange to yellow-green, soft, slightly slimy, thin to thick irregular masses up to 8 inches (20.3cm) across, frequently with projections; oscules often raised, conspicuous. In bays and open coast, central and southern California, intertidal to about 15 feet (5m).

Isociona lithophoenix: deep red, usually hard encrustations 1/8-inch (0.3cm) thick; occasionally more than 1 inch (2.5cm) thick. Hardest of all the red sponges. Central California to Baja California, intertidal to 50 feet (15m) or more.

Lissodendoryx noxiosa: yellow, fairly firm but spongy, up to 3 inches (7.6cm) thick, 6 inches (15cm) across; smooth but irregular surface, oscules poorly defined; odoriferous. Central and southern California, intertidal to 80 feet (24m).

Microciona parthena: red, moderately stiff, thin patches to masses 1 inch (2.5cm) thick; surface riddled with tiny crevices like a bread-crumb, oscules obscure. Southern California, 70 feet (21m) or deeper.

Ophalitaspongia pennata: scarlet to orange, velvety, thin broad sheets; small unraised oscules located at center of radiating grooves (more easily seen out of water). Northern and central California, intertidal to shallow.

Penares cortius: gray, firm; slippery surface; scattered oscules up to ¼-inch (0.6cm) in diameter. Central California to Baja California, intertidal to 50 feet (15m) (Plate 2e).

Plocamia karykina: scarlet, firm, broad sheets usually less than ½-inch (1.3cm) thick; surface sometimes laced with fine white filaments; small oscules, slightly raised. Red throughout; produces mucus copiously when injured. Central California to Baja California, intertidal to 60 feet (18m).

Prianos problematicus: drab white, firm clusters of urn-shaped growths with oscules up to ¼-inch (0.6cm) in diameter. Central California to Baja California, shallow subtidal to 50 feet (15m) or more.

Spheciospongia confoederata (Liver Sponge): gray, firm but flexible, up to 1 foot (30.5cm) thick and 15 feet (5m) long; often ridged with conspicuous oscules. Our largest sponge. Central California to Baja California, intertidal to 40 feet (12m).

Stellata clarella: dirty white, sometimes partially covered by a thick, gritty, gray skin; spongy, up to 3 inches (7.6cm) thick and 18 inches (46cm) across. Surface plush, with sharp spicules (beware!). (See group II, below, for *S. estrella*, a similar ovoid sponge.) Central and southern California, intertidal to 80 feet (24m) (Plate 1d).

Tedania topsenti: red, orange, or pink; soft, spongy, biscuit-sized; oscules conspicuous, often on ridges or papillae. Central and southern California, intertidal to 80 feet (24m).

Tedanione obscurata: bright vermillion; ½-inch (1.3cm) craterlike tubes project up through turf. Central California to Baja California, intertidal to 40 feet (12m).

Verongia thiona: brilliant yellow, spongy, up to 2 inches (5.1cm) thick, 5 inches (12.7cm) across; smooth, irregular papillated surface, oscules large but sparse. Turns dark after preservation. Southern California, intertidal to 80 feet (24m).

Xestospongia vanilla: pearly white to pale yellow, stony-hard, broad encrusting sheets up to ½-inch (1.3cm) thick; oscules often raised. Common underneath rocks and over-hangs; entire coast, intertidal to shallow subtidal.

II. TUBELIKE, OVOID, OR VASE-SHAPED INDIVIDUALS OR GROUPS

Leucetta losangelensis: white to pale brown groups of fairly soft tubes about 1 inch (2.5cm) long and ¼-inch (0.6cm) in diameter, the group being up to 5 inches (12.7cm) across. Central and southern California, intertidal to 80 feet (24m).

Leuconia heathi: dirty white, fairly soft, globular to pear-shaped, up to 4 inches (10cm) high by 5 inches (12.7cm) in diameter but usually much smaller; surface spiny. Dense band of spicules surrounds oscule (beware!). Central California to Baja California, intertidal to 240 feet (73m).

Leucosolenia eleanor: gray-white, bundles of soft anastomosing tubes ¼-inch (0.6cm) in diameter. *L. macleayi*: gray-white, soft vaselike to cube-shaped masses on short stalks up to 1 inch (2.5cm) tall. Central California to Baja California, intertidal to 40 feet (12m) or more.

Rhabdodermella nuttingi: white, fairly soft, vaselike individuals often in close association, ¼-inch (0.6cm) diameter by 1 inch (2.5cm) high. Central California to Baja California, intertidal to 80 feet (24m).

Stelleta estrella: dirty white, spongy, small ovoid knobs to cakes 2 inches (5.1cm) thick and 3 inches (7.6cm) across; surface spiny; accumulated detritus. (See group I, above, for *S. clarella*, a similar flattened sponge.) Southern California, intertidal to 100 feet (30.5m) (Plate 1d).

III. ANIMAL APPROXIMATELY SPHERICAL

Craniella (*Tetilla*) *arb*: young are tan, older drab; firm, up to 3 inches (7.6cm) in diameter; dense spicules surround oscule and cover surface. Central to southern California, intertidal to 100 feet (30.5m).

Geodia mesotriaena: young animals may approach spherical shape; see group I, above, for description of more common flattened form.

Tethya aurantia: yellow to orange exterior, spongy, up to 4 inches (10.2cm) in diameter; surface warty, oscules obscure. Brown interior. Central California to Baja California, intertidal to 100 feet (30.5m) (Plate 2b).

IV. CONSPICUOUS UPRIGHT LOBES, OR FINGERLIKE OR LEAFLIKE PROJECTIONS

Dysidea amblia: gray, spongy, forms a branching system of stalks about ½-inch (1.3cm) in diameter, 1 foot (30.5cm) or more high; surface papillate, oscules tiny. Occasional; central and southern California, intertidal to 70 feet (21m).

Esperiopsis rigida: bright red to yellow, firm, forms a branched structure of elongate clublike stalks with frayed tips; 1 foot (30.5cm) or more tall, smooth, occasional oscules. Stalks taper basally to cylindrical shapes ½-inch (1.3cm) in diameter. Northern California, 10 to 80 feet (3-24m).

Haliclona ecbasis: brown, firm, short club-shaped stalks; structure porous like bread. Central California to Baja California, intertidal to 50 feet (15m) or more.

Hemectyon hyale: orange to red, firm but flexible; smooth, oscules obscure. Forms a branched system of thin leaflike projections 1/8-inch (0.3cm) thick, 1 inch (2.5cm) wide, several inches (mm) long–entire animal may exceed 1 foot (30.5cm). Southern California and Baja California, 70 feet (21m) or deeper.

Polymastia pachymastia: yellow, firm but flexible, forms a conspicuous group of fingerlike protrusions from an encrusting basal sheet 8 inches (20.3cm) or more across. Projections ½-inch (1.3cm) in diameter, 2 inches (5.1cm) high; oscules on projection tips, smooth except for spicule plush on base. Central and northern California, intertidal to 50 feet (15m) (Plate 2b).

Toxadocia sp.: gray-white, firm, branching cylindrical stalks tapering toward base, about 1 foot (30.5cm) tall, 1 inch (2.5cm) or less diameter; occasional oscules, smooth. Central and southern California, deeper than 40 feet (12m) (Plate 2f).

THE PELAGIC COELENTERATES (PHYLUM)

The coelenterate classes Hydrozoa and Scyphozoa are well represented in the nearshore California plankton; indeed, the largest and most conspicuous of all our plankton are a heterogeneous group called jellyfishes. Unlike the harmless benthic coelenterates of California, several pelagic species sting severely, causing swelling, rashes, welts, and pain that may persist for days. Often a fragment of transparent coelenterate tissue remains undetected and clinging to the affected skin. If the fragment contains nematocysts (stinging cells), it will cause further damage. Rub the burned skin gently with wet sand immediately after contact; later apply a dilute ammonia solution.

CLASS PELAGIC HYDROZOA

Calycophora, a Suborder of the Order Siphonophora (several species: transparent, rigid, helmet-shaped parent produces long segmented chains that break loose and lead a free existence as drifting stages called eudoxomes. A eudoxome may be ½-inch (1.3cm) across and many feet (m) long; it is transparent or slightly yellow, and stings very painfully.

Polyorchis pacifica: transparent, size and shape of a large egg; margin of umbrella smooth, many tentacles, 4 faint radial canals with pink or red gonads beneath.

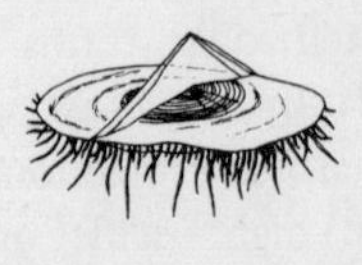

Velella lata (By-the-wind Sailor): transparent, stiff, oval, hollow; upper portion with conspicuous ridge (sail) supports blue, soft underportion and tentacles (actually a colony of many individuals). Floats on surface.

CLASS PELAGIC SCYPHOZOA

Aurelia: transparent; disc-shaped translucent bands (the gonads) seen when viewed from top; marginal tentacles small and numerous. *A. aurita*: 8-lobed margin. *A. labiata*: 16-lobed margin.

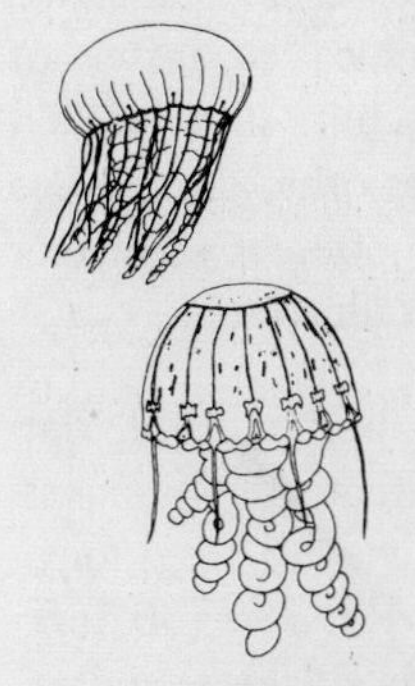

Chrysaora melanaster: radial red-brown lines, umbrella hemispherical, 24 marginal tentacles not in clusters, prominent lobes below mouth.

Pelagia panopyra: purple-brown radial markings, umbrella hemispherical, 8 evenly spaced marginal tentacles, prominent lobes below mouth (Plate 4d).

THE BENTHIC COELENTERATES (PHYLUM)

Limited space allowed us only to describe conspicuous genera and species of the hydroids, corals, and anemones. We will barely scratch the surface, however, and the reader with access to a dissecting microscope will discover a wealth of tiny fascinating coelenterates colonizing the majority of solid objects recovered from underwater. Many coelenterates exhibit an alteration of generations between an attached polyp form and a free-swimming medusa form. The conspicuous medusae (jellyfish) are treated under pelagic coelenterates, above, and we will confine ourselves here to the attached forms.

CLASS HYDROZOA

Common California hydrozoans include the hydroids and one coral-like genus. The animals are small to microscopic, but many form colonies and together build structures that are very conspicuous and useful for identification. Sometimes the colony must be collected and examined topside with a hand lens in order to classify the animal properly. Shape and position of the *theca* (the protective cup surrounding each animal of the colony) is often critical for identification.

I. HYDROZOANS RECOGNIZABLE UNDERWATER

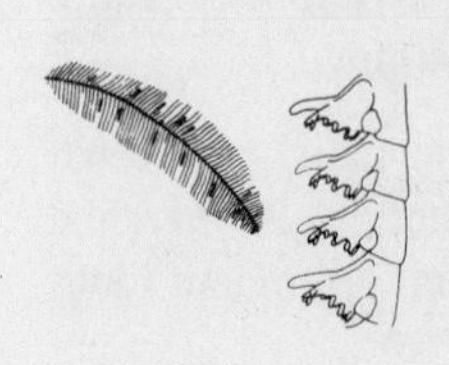

Aglaophenia (Ostrich Plume; several species): dark, flexible, dense branching; typical colony resembles a cluster of feathers. (Sparse branching in a feather-like structure indicates other groups–see below.) Intertidal to 80 feet (24m) or more. *A. struthenoides* illustrated.

Allopora porphyra (Hydrocoral): bright purple, hard, flat encrusting sheets or upright branched structures; branches ¼-inch (0.6cm) in diameter, 2 or more inches (5.1cm) long. Common on offshore pinnacles, down to 100 feet (30.5m), entire coast; intertidal in northern California (Plate 3a and c).

Corymorpha palma (Sea Palm): transparent, delicate without hard parts; stem up to 3 inches (7.6cm) tall, apex with 2 bands of tentacles. Arises from mud or fine sand, common in bays or estuaries; intertidal to 30 feet (9m).

Tubularia (several species): stem up to 3 inches (7.6cm) tall in protected strawlike tube; hydranth (head) has basal whorl of long tentacles, cluster of short tentacles at apex. Some species solitary, others colonial. Entire coast, intertidal to more than 100 feet (30.5m).

II. HYDROZOANS REQUIRING TOPSIDE EXAMINATION FOR IDENTIFICATION

Theca stalked, resembles a wine glass; complete identification requires microscopic study of sexual stages; common genera are:
Campanularia
Obelia

Theca not stalked

Theca unilateral

Theca bilateral

Theca margin smooth, colony slender and flexible: *Plumularia*

Theca toothed and sculptured, colony stiff, featherlike: *Aglaophenia*

Thecae opposite each other: *Sertularia*

Thecae alternate or subalternate

Theca margin smooth, colony plumelike: *Abietinaria*

Theca margin toothed, colony with few or no branches: *Sertularella*

CLASS SCYPHOZOA

Stauromedusae (several genera): body wineglass-shaped, surmounted by 8 lobes tipped with round tentacle clusters. Usually attached to seaweeds, and assumes color of substrate.

CLASS ANTHOZOA

I. Rocky-bottom anemones
II. Sandy-bottom anemones
III. Soft corals
IV. Stony corals
V. Colonial anthozoans

I. ROCKY-BOTTOM ANEMONES

Anthopleura: numerous pointed tentacles, column rough with tubercles holding sand, shell particles, etc.; numerous small spherules just beneath tentacles. *A. aggregata* (*elegantissima*): tubercles in longitudinal rows; aggregates in groups. *A. artemesia*: lower two-thirds of column white to pink, tentacles bright colors; solitary. *A. xanthogrammica*: tubercles arranged irregularly; solitary. Entire coast, all intertidal; *A. elegantissima* shallow, others to 100 feet (30.5m) (Plate 3a).

Corynactis californica: various colors; numerous tentacles terminating in knobs, column smooth, about ½-inch (1.3cm) in diameter by ½-inch (1.3cm) tall; aggregating. Central California to Baja California, intertidal to more than 100 feet (30.5m) (Plate 3a, b, and c).

Diadumene (formerly *Sagartia*): various shades of green; many pointed tentacles, column smooth and often striped, up to ½-inch (1.3cm) tall. Entire coast, intertidal to shallow subtidal.

Epiactis prolifera: brown to red, many pointed tentacles, column smooth, white stripes at base; about ½-inch (1.3cm) tall, up to 1 inch (2.5cm) in diameter; young anemones often attached on column. Entire coast, intertidal to 30 feet (9m).

Metridium senile: usually white, occasionally brown or orange, 1 foot (30.5cm) or more tall, 4 inches (10.2cm) in diameter; short numerous tentacles form fuzzy frilled border. Entire coast, intertidal to more than 100 feet (30.5m) (always deep when south of Point Conception) (Plate 8e).

Paracyathus stearnsi: a small, cream-colored species, forms colonies on gorgonian corals, killing the coral and leaving the skeleton to support the invader. Southern California and Baja California, 40 to more than 100 feet (12-30.5m).

Tealia: column red or red with green patches, many pointed tentacles. *T. coriacea*: many tubercles on column holding sand, shell, etc. *T. crassicornis*: few or no tubercles. *T. lofotensis*: a few white tubercles, arranged longitudinally in rows (Plate 4b). Entire coast, intertidal to more than 100 feet (30.5m); only *T. coriacea* south of the Channel Islands, confined to cold waters.

II. SANDY-BOTTOM ANEMONES

Edwardsiella californica: 16 tentacles, mouth slightly raised above disk, column three-quarters covered with brown epidermis and displays 8 ridges; about ¼-inch (0.6cm) in diameter, 4

a. *Acarnus erithacus, Synoicum* sp. (white nodules at lower center). By R. Hollis.

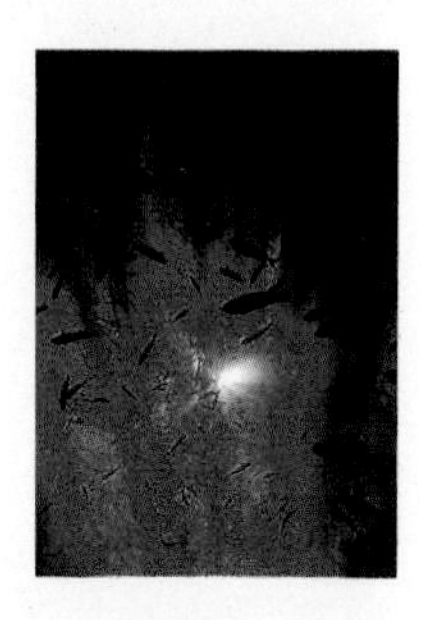

b. Submarine forest of *Macrocystis pyrifera.* By C. Nicklin.

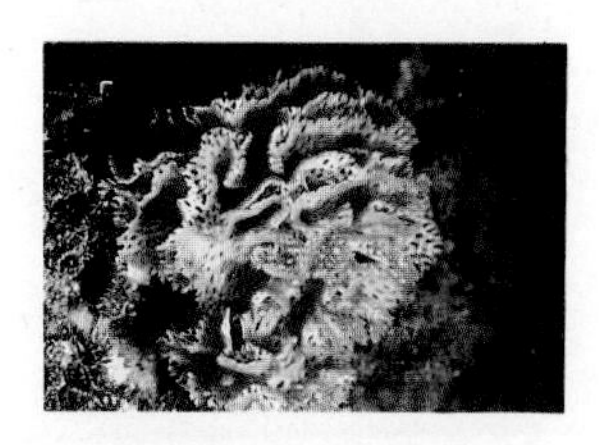

c. *Phidolopora pacifica.* By C. Nicklin.

d. *Stelleta* sp. By R. Hollis.

e. *Parastichopus* sp., *Thalamoporella californica,* and *Calliarthron cheilosporioides.*

PLATE 1.

a. *Hopkinsia rosacea.*

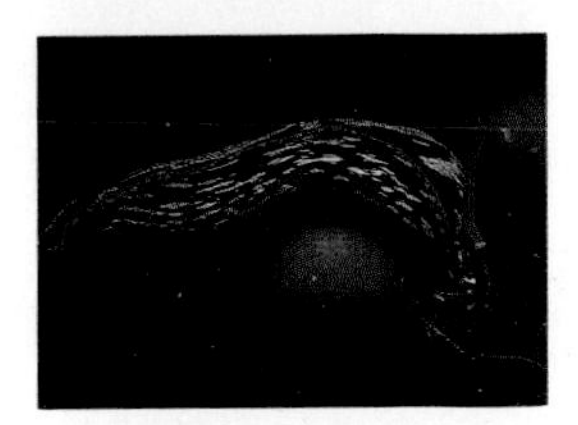

c. *Chelidonura inermis.* By R. Bergero.

d. *Sebastes chrysomelas.* By. C. Nicklin.

e. *Gymnothorax mordax, Penares cortius,* and *Hippolysmata californica.* By C. Nicklin.

b. *Tethya aurantia, Polymastia pachymastia, Corynactis californica,* and *Ballanophyllia elegans.* By. R. Hollis.

f. *Toxodocia* sp., *Henricia leviuscula,* and *Strongylocentrotus franciscanus.* By C. Nicklin.

PLATE 2.

a. *Anthopleura xanthogrammica, Corynactis californica* and *Allopora porphyra.* By R. Hollis.

b. *Eudistylia polymorpha,* and *Corynactis californica.* By C. Nicklin.

c. *Allopora porphyra,* and *Corynactis californica.* By R. Hollis.

PLATE 3.

a. Cerianthid anemone, and *Ophiothrix spiculata*. By C. Nicklin.

b. *Tealia lofotensis*. By R. Hollis.

c. *Astrangia lajollaensis*. By C. Nicklin.

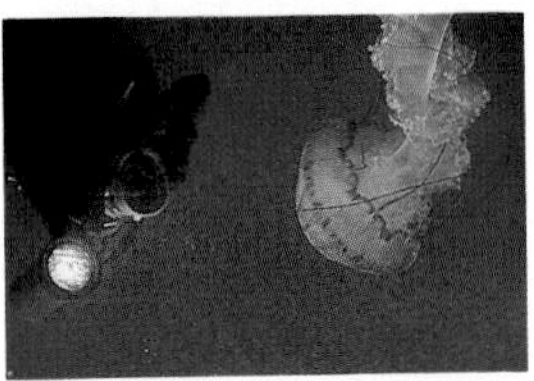

d. *Pelagia panopyra*. By C. Nicklin.

e. Juvenile *Hypsypops rubicunda*. By C. Nicklin.

f. *Lima hemphilli*. By R. Hollis.

PLATE 4.

a. *Duvecelia festiva* on *Lophogorgia chilensis.* By C. Nicklin.

b. *Phidiana nigra.*

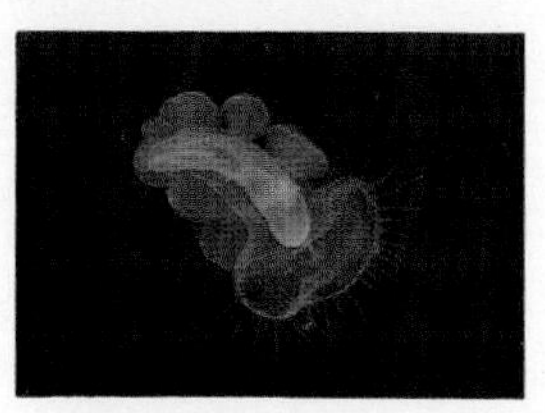

c. *Chioraera leonina.*

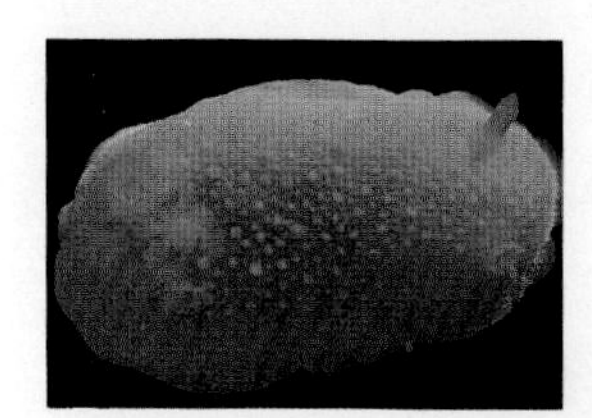

d. *Austrodoris odhneri.*

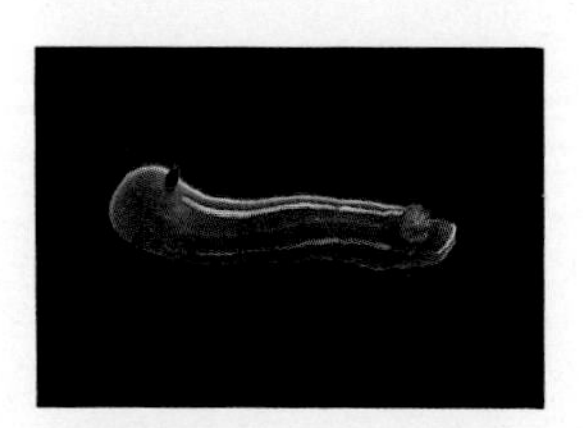

e. *Glossodoris porterae.*

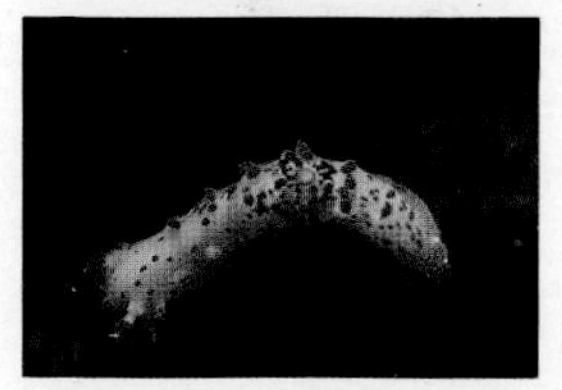

f. *Triopha carpenteri.* By R. Bergero.

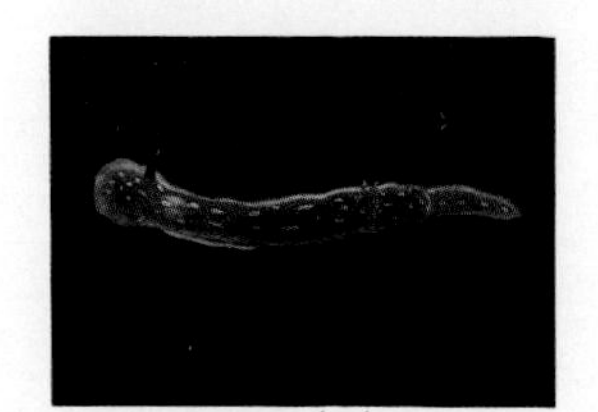

g. *Glossodoris californiensis.*

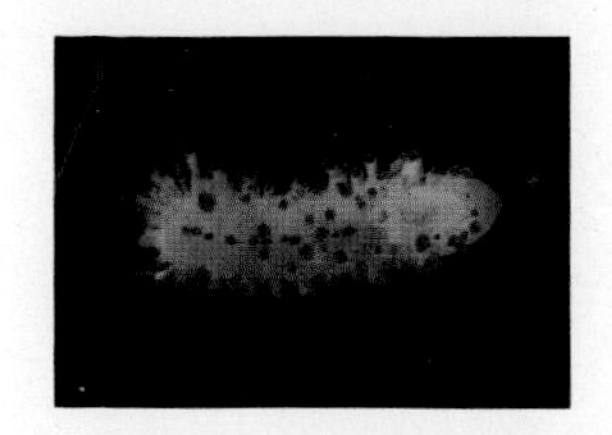

h. *Laila cockerelli.*

PLATE 5.

a. *Archidoris montereyensis.*

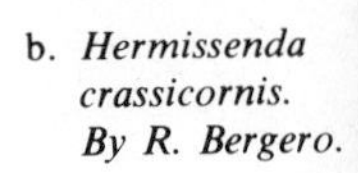

b. *Hermissenda crassicornis.* *By R. Bergero.*

c. *Cadlina limbaughi.*

d. *Flabellinopsis iodinea.* By R. Bergero.

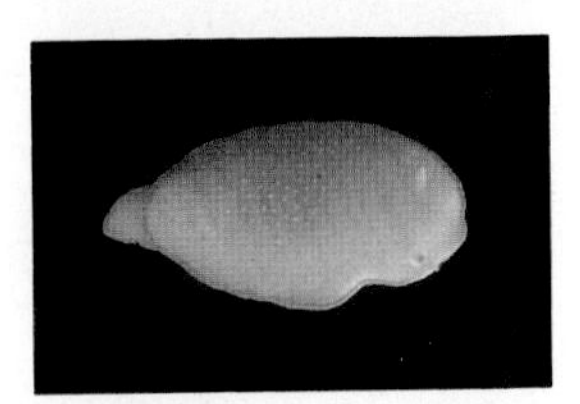

e. *Cadlina marginata.*

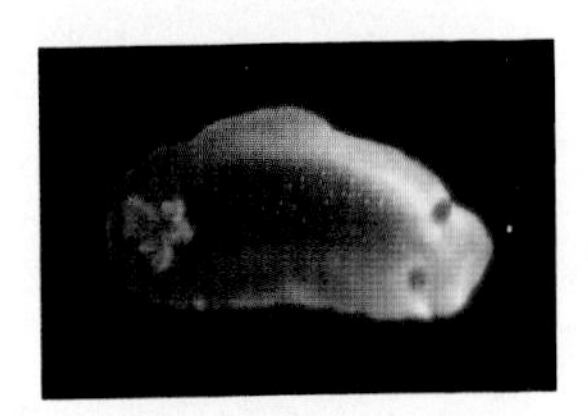

f. Dendrodoris albopunctata

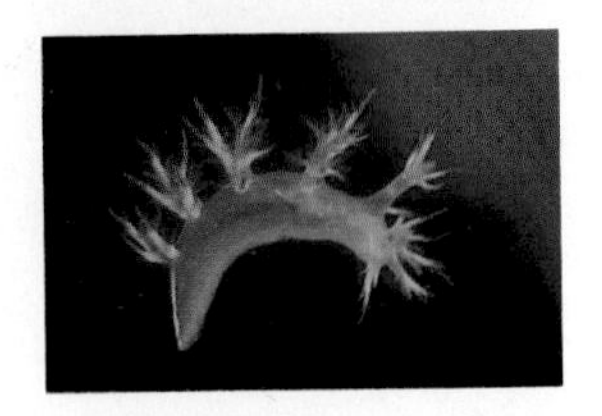

g. *Dendronotus frondosus.*

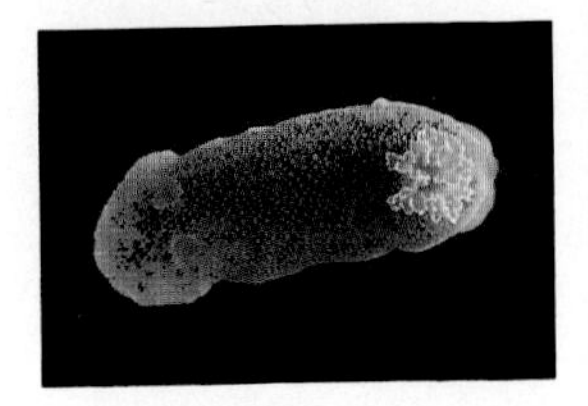

h. *Anisodoris nobilis.*

PLATE 6.

. *Pisaster giganteus, Corynactis californica,* on shell of *Balanus nubilis.* By R. Hollis.

b. *Pychopodia helianthoides, Patiria miniata.* By R. Hollis.

c. *Strongylocentrotus franciscanus, Balanophyllia elegans, Didemnum carnulentum, Dodecaceria* sp. By R. Hollis.

PLATE 7.

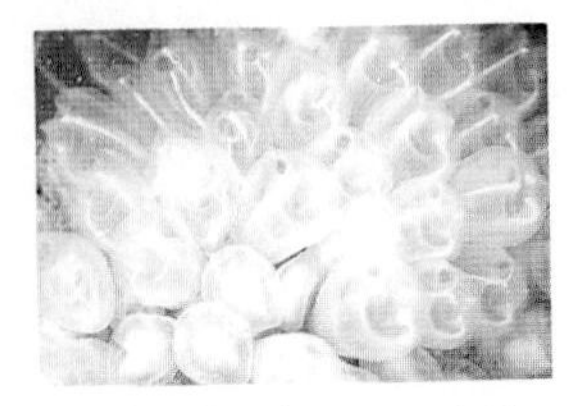

a. *Clavelina huntsmani*. By R. Bergero.

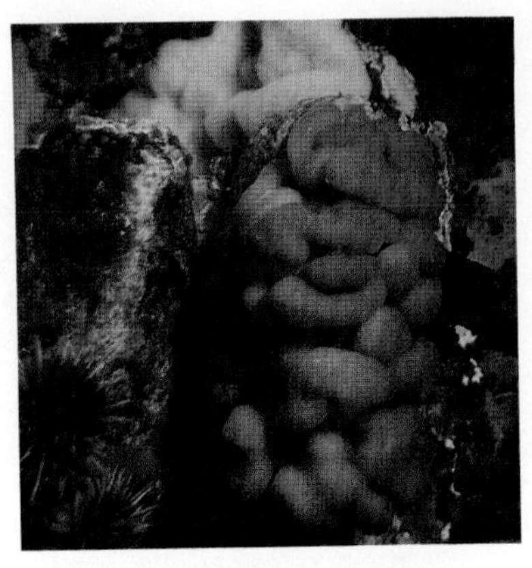

b. Colonial tunicates, probably *Cystodites* sp., and *Amaroucium* sp. *Peyssonelia* sp., *Strongylocentrotus purpuratus*. By R. Hollis.

c. *Zalophus californianus*, *Phyllospadix torreyi*. By C. Nicklin.

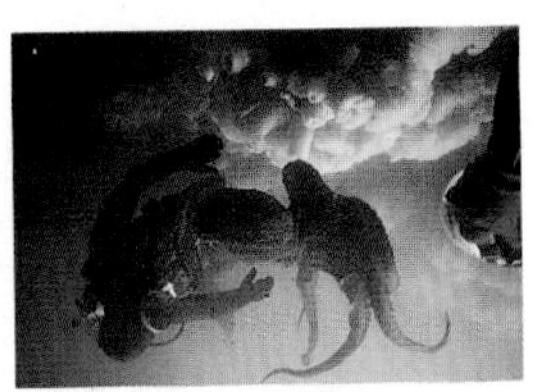

d. *Octopus dofleini*. By C. Nicklin.

e. *Metridium senile*. By R. Hollis.

f. *Loligo opalescens*. By C. Nicklin.

PLATE 8.

inches (10cm) long, anchors with bulbous base. Intertidal to 30 feet (9m), southern California and Baja California.

Harenactis attenuata: 24 tentacles, often banded; column whitish, up to 1 inch (2.5cm) in diameter, more than 1 foot (30.5cm) long, anchors by a bulbous base. Southern California and possibly entire state, intertidal and shallow subtidal.

Tube anemones (order Ceriantheria, several species): tentacles in 2 concentric whorls, up to 2 inches (5.1cm) in diameter, 6 inches (15cm) long. Most conspicuous is *Pachycerianthus*, occurring from 10 to more than 100 feet (3-30.5m), southern California and Baja California (Plate 4a).

III. SOFT CORALS

Order Alcyonacea

Eunephthya rubiformis (Sea Strawberry): bright red, soft lobes with scattered white polyps, 8 tentacles per polyp; up to 4 inches (10cm) tall. Trinidad Bay northward, shallow subtidal to about 20 feet (6m).

Order Gorgonacea–Gorgonian Corals, the Sea Fans

Adelogorgia phyllosclera: orange-red with yellow polyps; sparse branching in one plane. Color blackens on drying. Southern California, 120 feet (36.5m) or deeper.

Eugorgia rubens: red-purple to violet with white polyps, up to 6 feet (2m) tall but usually 1 foot (30.5cm) or shorter; tortuous branching in one plane, forming reticulate pattern. Southern California and Baja California, 90 feet (27m) and deeper.

Lophogorgia chilensis: salmon pink with white polyps, 3 feet (1m) tall; branching sparse and in all directions. Southern California and Baja California, 50 feet (15m) and deeper (Plate 5a).

Muricea: rusty brown, up to 3 feet (1m) tall, very flexible; dense branching more or less in one plane. *M. californica*:

yellow polyps; *M. fructicosa*: white polyps. Southern California and Baja California, 20 to 50 feet (6-15m).

IV. STONY CORALS

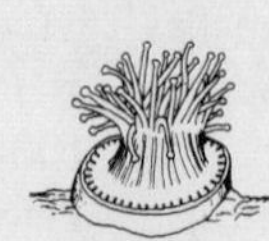

Astrangia lajollaensis: yellow orange, or red polyp, in semicylindrical stony shell (coralite); ½-inch (1.3cm) in diameter, up to 4 inches (10.2cm) tall; often aggregates in groups. Inside of coralite partitioned by many radial walls (*septa*). Southern California and Baja California, especially Coronados and San Clemente islands, 40 feet (12m) or deeper (Plate 4c).

Balanophyllia elegans: yellow or orange polyp in cup-shaped coralite; always solitary. Radiating septa of coralite form starlike pattern. Entire coast, in cold water, intertidal to 80 feet (24m) or more (Plates 2b and 7c).

V. COLONIAL ANTHOZOANS

Acanthoptilum gracile (Sea Pen): translucent white stalk (*axis*) projects from sand up to 2 feet (61cm); fluffy lateral lobes give featherlike appearance, main axis smooth and flexible. Central California, 10 to 300 feet (3-91.5m).

Leioptilus guerneyi (Pink Sea Pen): pale pink, soft axis bearing leaflike lobes in a stacked arrangement projects from coarse sand; buried portion elongate, bulbous. Entire coast, 90 feet (27m) and deeper.

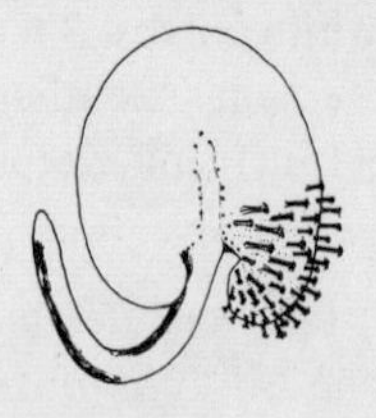

Renilla kollikeri (Sea Pansy): purple, heart-shaped disk with stubby tail, upper side a profusion of transparent polyps; lies slightly buried at sand surface. Southern California and Baja California, intertidal to 30 feet (9m).

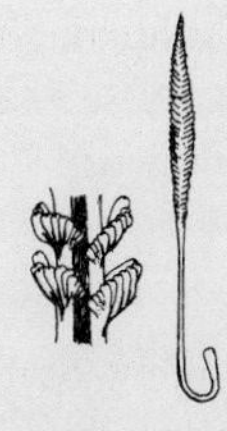

Stylatula elongata (White Sea Pen): translucent white axis adorned with fluffy lateral lobs projects up to 2 feet (61cm) from sand; axis brittle and rough to the touch, due to serrations. Central California to Baja California, intertidal to 200 feet (61m).

PHYLUM CTENOPHORA

Ctenophores are delicate pelagic creatures somewhat resembling coelenterate jellyfishes. Most of the body is transparent tissue with a jellylike consistency. Tentacles and lobes may be present. Locomotion is provided by cilia arranged symmetrically on the outside surface in 8 bands; the cilia are microscopic, but cause iridescence in glancing sunlight.

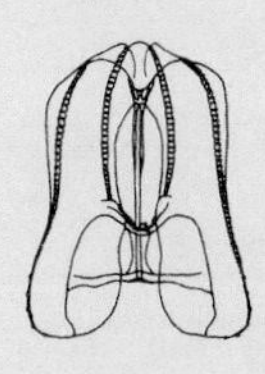

Bolinopsis: extremely transparent, flattened body, often only reddish bands of cilia visible; 12 inches (30.5cm) or more tall, large lobes hanging downward produce highly convoluted shape. Disintegrates when handled. Entire coast, open water.

Pleurobrachia bachei: ovoid to spherical, up to ½-inch (1.3cm) in diameter; often congregating near surface. Entire coast.

PHYLUM BRYOZOA or ECTOPROCTA

Individually, bryozoans are microscopic, but collectively the colonies are often conspicuous and beautiful. Coloration is typically subdued. Bryozoans are usually responsible for the delicate pastel backgrounds in artistic photographs of the sea floor. Bryozoan colonies may resemble thin encrustations by plants, sponges, or tunicates, or the erect, branching forms may be mistaken for small plants or hydroids. Close examination, however, reveals a fine reticulated network in the bryozoan colony, caused by hundreds of tiny compartments (*zooecia*) densely packed side by side. Each zooecium contains one bryozoan polyp and is visible to the naked eye, or can easily be distinguished with a hand lens. Hydroids are also compartmentalized, and this may cause confusion. Arborescent bryozoan skeletons tend to be calcified, making them hard and brittle, especially after drying; hydroid skeletons are always flexible, and the compartmentalization is usually too small to be seen clearly with the naked eye.

With a little experience, identification to genus is often possible in the field with a hand lens. Further classification to species usually requires a microscope, unless the genus contains only one species or is dominated primarily by a species.

MAJOR SUBDIVISIONS BY GROSS APPEARANCE

I. Flat, encrusting sheets

II. Thin laminae, arising from a flat sheet or base

III. Upright stalks, often delicately branched

IV. Other appearance

I. FLAT ENCRUSTING SHEETS

The term "flat" is used to mean that the encrustation generally assumes the shape of the surface underneath and does not add substantial irregularities. If the substrate is bumpy, a flat encrustation will not lie in one plane. Some bryozoan colonies initially form a small flat encrustation, and stalks (*laminae*) develop subsequently. If a flat specimen is small, therefore, it may actually belong in one of the other groupings. Use the largest colony available for identifying the flat forms.

Cauloramphus spiniferum: zooecium oval, close-packed, rimmed with spines, giving the colony a velvet-like appearance and feel. Entire coast, intertidal to more than 40 feet (12m).

Cryptosula pallasiana: zooecium many-sided, like a partially crushed box; peppered with many large pores, aperture with rim; occasionally produces laminae. Entire coast, intertidal to 180 feet (55m).

Hippothoa hyalina: zooecium cocoon-shaped with conspicuous transverse ribs, no pores; colony surface often glistening, chalky white. Entire coast, intertidal to more than 50 feet (15m).

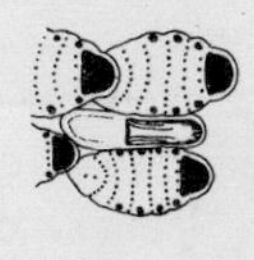

Lyrula hippocrepis: zooecium cocoon-shaped, pores in transverse rows; one border of aperture straight, topped by "inverted lyre"; occasionally forms laminae. Entire coast, intertidal to 450 feet (137m).

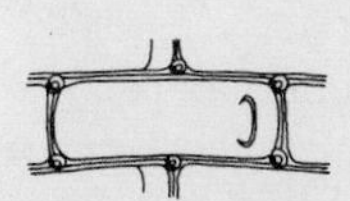

Membranipora (several species): zooecium commonly a rectangular box; close-packed, colonies whitish to dirty yellow. Abundant on kelp and other seaweeds; entire coast, intertidal to 200 feet (61m). *M. membranacea* illustrated.

Microporella (several species): zooecium cocoon-shaped, many pores, spines around aperture, often a bonnet-shaped ovicell above aperture. On stones and shells; entire coast, intertidal to 450 feet (137m). *M. californica* illustrated.

Rhyncozoon rostratum: young cooecium broadly oval; tubercles around aperture; colonies cream or yellow to orange, older colonies appear as deep apertures scattered randomly over a highly irregular surface built up by secondary calcification. Very common on stones and shells; entire coast, intertidal to 600 feet (183m).

II. THIN LAMINAE ARISING FROM FLAT BASE

Cryptosula and *Lyrula* occasionally form laminae, but are described in group I, above.

Flustrella corniculata: oval zooecium concealed by spiny plush; spines usually with 4 prongs, occasionally 6; colony soft, flexible, tan to dark brown. Northern to southern California, intertidal to 200 feet (61m).

Hippodiplosia insculpta: zooecia with parallel walls, close-packed in rows, occur on both sides of lamina; laminae often frills or fan-shaped, yellow to orange. Common, entire coast, intertidal to 700 feet (213.5m).

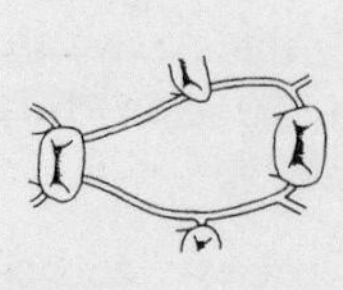

Pherusella brevituba: zooecium cocoon-shaped, aperture a flaring tube; laminae light brown, flexible, and leathery, often fan-shaped. Common in seaweed hold-fasts and lower blades; southern California and Baja California, intertidal to 50 feet (15m).

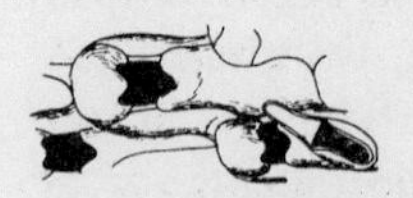

Phidolopora pacifica: zooecium roughly oval, complex aperture; laminae convoluted and filled with holes, anastomose frequently to form an intricate rigid structure; orange to rose-pink. Entire coast, intertidal to 600 feet (183m) (Plate 1c).

III. UPRIGHT STALKS, OFTEN DELICATELY BRANCHED

Bugula (several species): zooecium irregular shape but elongate; dense flexible branching, purple to yellow to brown, branches all curve inward; colonies are small "brushes" up to 6 inches (15.2cm) tall. *B. neritina* (illustrated) lacks spines (distinguishes from *Scrupocellaria*, below). Common on floats, docks, jetties, etc.; entire coast, intertidal to 240 feet (73m).

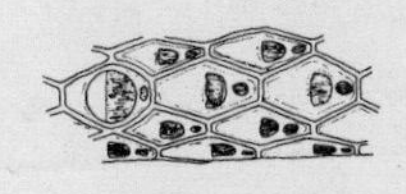

Cellaria mandibulata: zooecium coffin-shaped; moderate to dense branching, long internodes joined by dark band at nodes; colony 1 to 3 inches (2.5-7.6cm) tall. Entire coast, intertidal to 500 feet (152m).

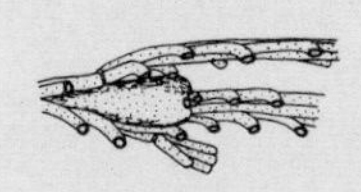

Crisia (several species): zooecium a long tube; branches composed of two alternating zooecia in rows, 3 or more zooecia per internode; branches are thin and fragile, but colony may be dense whitish tuft 1 or 2 inches (2.5-5.1cm) high. Entire coast, intertidal to 400 feet (122m). *C. maxima* illustrated.

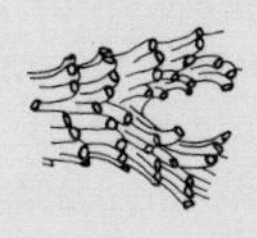

Diaporoecia californica: zooecium tubular, arranged in rows, sometimes embedded in a matrix; branches rigid, anastomosing, blunt or clublike at tips; colony yellow to orange, up to 4 inches

(10.2cm) tall, resembles a miniature coral. Entire coast, intertidal to 600 feet (183m).

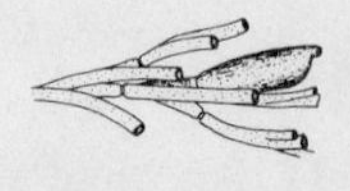

Filicrisia (two species): resembles *Crisia* (above), except only 1 to 3 zooecia per internode. *F. franciscana* (illustrated) and *F. geniculata* differ only in microscopic features of reproductive parts. Entire coast, intertidal to 150 feet (45.5m).

Flustrella corniculata: colony occasionally a cylindrical clublike stalk, more commonly leaflike as described in group II above.

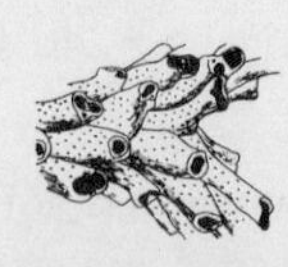

Lagenipora punctulata: zooecium flask-shaped; branching, cylindrical stems; colony rigid. May be confused as a small *Diaporoecia* (see above); zooecium is entirely different, however. Entire coast, intertidal to 600 feet (183m).

Scrupocellaria (several species): dense flexible branches; stems formed of two rows of roughly cylindrical zooecia, densely covered with spines and other projections. *S. diegensis* (illustrated) colonies 2 inches (5.1cm) or more high, orange. Common, entire coast, intertidal to 40 feet (12m).

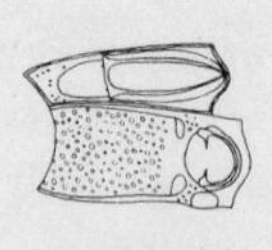

Thalamoporella californica: zooecium roughly rectangular with parallel walls; colony arises from encrusting sheet, forms dense whitish arborescent branches with long cylindrical stems tapering at internodes. Common under thick kelp canopies; southern California and Baja California, intertidal to 300 feet (91m) (Plate 1e).

IV. OTHER APPEARANCE

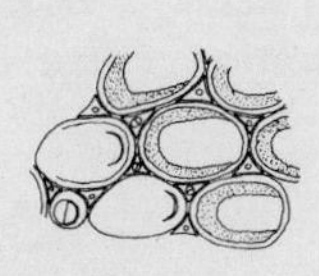

Antropora tincta: zooecia oval, disorientedly arranged; young portions of colony white, blending into pinkish-brown older areas. Typically encrusts old snailshells, producing a knobby appearance. Southern California and Baja California, shallow subtidal to 450 feet (137m).

Costazia robertsoniae (commonest of several *Costazia* species): zooecium roughly round; considerable distortion from secondary calcification (like *Rhyncozoon* in group I, above), but large pores usually conspicuous. Forms nodules, occasionally branches and encrusting sheets, on stems of coralline algae. Entire coast, intertidal to 350 feet (107m).

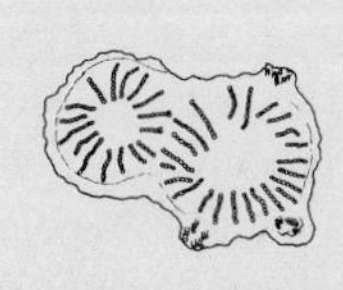

Disporella californica: tubular zooecia arranged in radiating rows; colony round, dome-shaped, blue, up to ½-inch (1.3cm) in diameter. A subtropical species; encrusting rocks, most commonly around islands, southern California, 30 to 400 feet (9-122m).

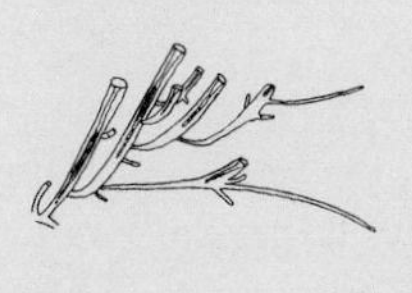

Victorella: polyp completely hidden within matrix of fine sedimentary grains; colony resembles a clump or sheet of mud adhering to stalks, shells, or stone. Strong squeezing with tweezers forces polyps out. Southern California, shallow to 60 feet (18m) or deeper. *V. pauida* illustrated.

THE SEGMENTED WORMS (PHYLUM ANNELIDA)

Unlike the more primitive worms, an annelid is divided lengthwise into more or less equivalent compartments or segments. Usually the segmentation extends to surface structures, so that the animal appears to be composed of many similar sections stuck together. The segments display a variety of organs or structures useful for identification, often including:

lobes extending laterally (*parapodia*)
bristles from the parapodia (*setae*)
slender projections from the parapodia (*cirri*)
branching projections from the parapodia (*branchiae* or *gills*)
scales or platelets on the upper surface (*elytra*)

Anteriorly the first two segments are modified to form the head, and may display large fleshy protuberances (*palps*) or more slender branched or unbranched tentacles. A small hard circular structure (*operculum*) sometimes serves as a "door" for the tube. Eyes and jaws may or may not be present. The tail segments are also usually modified; sometimes it is difficult to distinguish head from tail.

All of the common annelids in California waters belong to the class Polychaeta. Polychaete worms occur in nearly every environment and frequently dominate in the shallow sediments. California boasts more than 600 species, and classification often requires painstaking scrutiny under a microscope. A few species can be easily recognized in the field because of unique appearances or habitats.

MAJOR SUBDIVISIONS BY HABITAT OR APPEARANCE

I. Protective tube
II. Unique habitat
III. Unique appearance

I. PROTECTIVE TUBE

A. Tube walls sand or lime, firmly adjoined, forming massive colony
B. Tube calcified, hard, encrusting, solitary or accidentally overlapping

C. Tube soft, parchment-like, solitary or loose clusters, few adhering particles

D. Tube solitary, partly or completely formed of particles or debris

A. Tube Walls Sand or Lime

Dodecaceria (several species): animal dark or black; no operculum, aperture 1/32-inch (.8mm) in diameter, grooved palpi, several paired tentacular cirri. Entire coast, intertidal to shallow subtidal (Plate 7c).

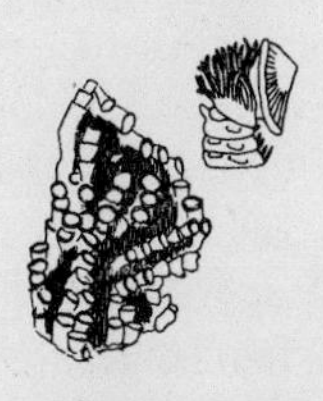

Phragmatopoma californica: operculum black, conical; many filamentous tentacles behind operculum; tubes of cemented sand grains. Common on kelp holdfasts, central California to Baja California, intertidal to 240 feet (73m).

Sabellaria cementarium: operculum amber, dome-shaped; filamentous tentacles behind operculum; tubes of cemented sand grains. Central California to Baja California, intertidal to 60 feet (18m).

Salmacina tribranchiata: no operculum; gills bright red, extend from tiny–1/32-inch (.8mm) in diameter–aperture; tubes white, fragile, up to 3 inches (7.6cm) long, sometimes in clusters several feet (m) across. Entire coast, intertidal to 100 feet (30.5m).

B. Tube Calcified, Solitary or Accidentally Overlapping

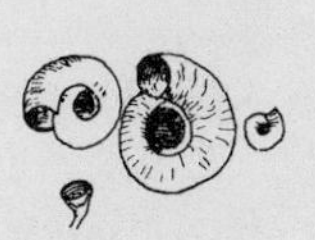

Dexiospira spirillum: white tube coiled tightly clockwise, coil about 1/8-inch (.3cm) in diameter. Encrusts seaweed blades and stones; entire coast, intertidal to 80 feet (24.5m).

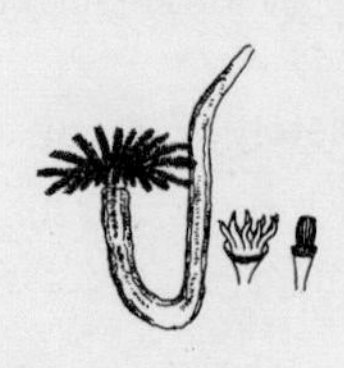

Eupomatus (several species): funnel-shaped operculum encircles tentacles, which are in single whorl, dull-colored; tube straight or moderately convoluted, about 1/16-inch (1.5mm) in diameter. Encrusts beneath stones and shells; entire coast, intertidal to shallow subtidal. *Eupomatus* sp. illustrated.

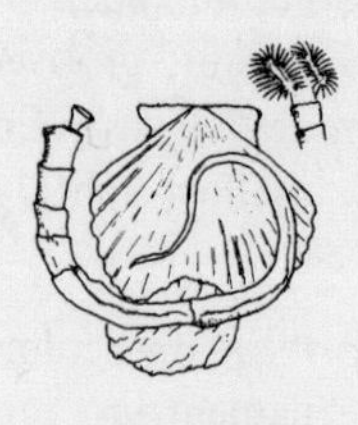

Serpula (several species; *S. vermicularis* [illustrated] common): operculum funnel-shaped, tentacles in 2 whorls, white to bright red; tube straight to convoluted, ¼-inch (0.6cm) in diameter, tapering. Encrusts reefs; entire coast, intertidal to 80 feet (24.5m).

Spirobranchis spinosus: funnel-shaped operculum, tentacles variously colored in 3 concentric whorls; tube ridged, producing median projecting spine on rim of aperture. Central California to Baja California, intertidal to 20 feet (6m).

C. Tube Soft, Solitary or Loose Clusters

Chaetopterus (several species): body translucent white, parapodia highly modified in various segments for different functions; tube double-ended, often U-shaped, tapering to apertures. In soft sediment or among rock crevices; entire coast, intertidal to 60 feet (18m). *C. variopedatus* illustrated.

Eudistylia: variously colored tentacles arranged in spirals, about 2 inches (5.1cm) across when extended; resembles miniature feather-duster. *E. polymorpha* (illustrated) in rocks, *E. vancouveri* in sediments; entire coast, intertidal to 60 feet (18m) or more (Plate 3b).

Phyllochaetopterus prolifica: animal 2 to 3 inches (5.1-7.6cm) long, 1/16-inch (1.5m) in diameter; tube double-ended, many inches (cm) long, bends frequently, irregularly. Entire coast, intertidal to 100 feet (30.5m).

Platynereis bicanaliculata: 2 short palps, numerous parapodia and setae resembling paddles; forms whitish mucus tubes in seaweed clumps. Entire coast, intertidal to 20 feet (6m).

Telepsavus costarum: about 1/16-inch (1.5mm) in diameter, up to 20 inches (50.8cm) long; tubes translucent to opaque, 24 inches (61cm) long by 1/8-inch (0.3cm) in diameter, bearing distinctive transverse ridges or annulations. Entire coast, intertidal to 60 feet (18m).

D. Tube Solitary, Partly or Completely Formed of Particles

Cistenides brevicoma: elongate curved conical tube open at both ends, formed of coarse sand grains; head spines blunt. Resembles *Pectinaria* (below). Entire coast, shallow subtidal to 40 feet (12m).

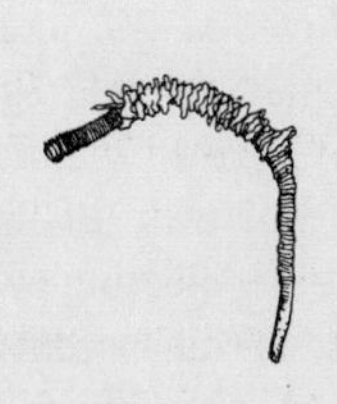

Diopatra ornata: 5 large and 2 small anterior tentacles, conspicuous lateral branchiae; tube partially hooded, thick-walled, membranous, bluish, covered with debris, projecting up to 3 inches (7.6cm) from sediments, becoming thin and bare beneath. Central California to Baja California, intertidal to 200 feet (61m).

Pectinaria californiensis: tube elongate, slightly curved, conical, open at both ends, yellowish, formed of very fine sediment grains; head spines tapering. Resembles *Cistenides* (above). Entire coast, intertidal to 600 feet (183m).

II. UNIQUE HABITAT

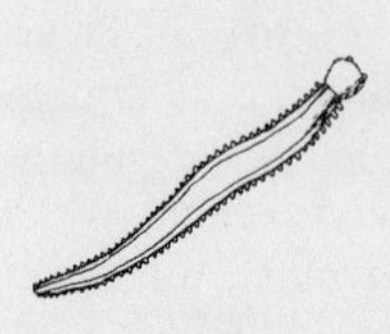

Flabelligera commensalis: many setal tufts along lateral margins; head bulblike, concealed by short setae. Lives among spines of *Strongylocentrotus* urchins; swims violently when disturbed. Southern California and Baja California, intertidal to 50 feet (15m).

Ophiodromis (*Podarke*) *pugettensis*: 3 tentacles and several palps on head, reddish-purple to brown or black, 1½ inches (3.8cm) long by ¼-inch (0.6cm) wide. Lives in grooves on underside of *Patiria* (sea star), also under stones and in surf grass. Entire coast, intertidal to 120 feet (36.5m).

Polydora (several species): 2 large anterior palps, branchiae commence after fifth segment. *P. websteri* drills tubes in mollusk shells; *P. paucibranchiata* builds a tube of mud particles. *P. commensalis* lives with hermit crabs; *P. armata* on coralline algae. *P. ligni* and *P. limicola* foul ship bottoms.

III. UNIQUE APPEARANCE

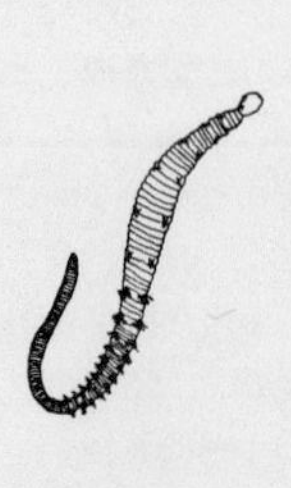

Arenicola (two species): black, conspicuously segmented, up to 20 inches (50.8cm) long; branchiae in midregion only; anterior region of greatest diameter, tapering posteriorly. Burrows in mid, sand; forms U-shaped tube, ejecting cylindrical castings. Egg-case is transparent sac with one end buried. *A. brasiliensis* (11 branchial pairs), southern California to San Francisco; *A. cristata* (illustrated) (16-18 branchial pairs) entire coast.

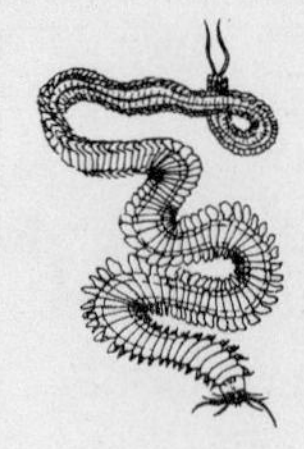

Nereis (several species): iridescent greenish-brown, conspicuously segmented; two palps and several tentacles anteriorly, parapodia modified for swimming, very little tapering at ends, up to 1

foot (30.5m) long; active. Entire coast, intertidal to 120 feet (36.5m). *N. vireus* illustrated.

SCALE WORMS (FAMILY)

Scaleworms are short broad animals with conspicuous elytra on upper surface. Classification rests on numbers of elytra pairs:

ELYTRA PAIRS	GENUS	LIVING HABIT	SPECIES
12	*Lepidonotus*		
15	*Harmothöe*	free-living	*H. imbricata*
15	*Harmothöe*	commensal on sea-cucumbers	*H. lunulata*
15	*Hesperonöe*	commensal with *Urechis*	*H. adventor*
15	*Hesperonöe*	commensal with *Callianassa*	*H. complanata*
18	*Halosydna*	commensal with *Thelepus*, free-living	*H. brevisetosa*
18	*Halosydna*	free-living	*H. johnsoni*
20+	*Arctonöe*	free-living and commensal	
20+	*Lepidothenia*	commensal with other polychetes	

Lepidonotus

Har. lunulata

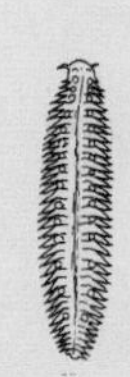

Hes. complanata

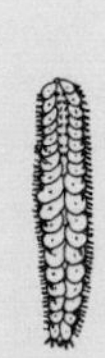

Hal. johnsoni

PHYLUM BRACHIOPODA

The brachiopods, or lamp shells, closely resemble bivalve mollusks and are often mistakenly identified as clams or mussels. Fortunately there is only one common species in California waters, and it is easily recognized—so that once the identity is learned, it is reasonable to assume that other bivalves are mollusks.

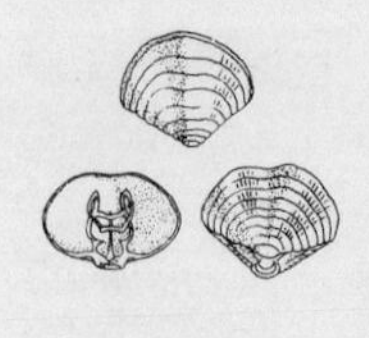

Terebratalia transversa: reddish shell with radiating ridges; one shell overlaps the other at base, upper junction of shells forming convoluted line; up to 2 inches (5.1cm) across, 1 inch (2.5cm) thick. Attaches to rock surfaces, entire coast, intertidal to 600 feet (183m); prefers cold water.

THE ECHIUROID WORMS

This small group of curious animals is occasionally very abundant at depths below the ordinary diving range. One species inhabits bay muds and sands in sufficient numbers to deserve attention.

Urechis caupo: flesh-colored, cigar-shaped, up to 20 inches (50.8cm) long but usually about 6 inches (15.2cm), short bristles at both ends; peristaltic body contractions pump water through the burrow, a U-shaped tube with surface openings 16 to 30 inches (40.6-76.2cm) apart, diameter less than ½-inch (1.3cm). Entire coast, intertidal to about 20 feet (6m) in cold bays.

Chapter 10

MARINE ANIMALS: MOLLUSCA

THE BENTHIC MOLLUSKS

Most mollusks have conspicuous calcareous shells–exceptions in California waters are nudibranchs, sea hares, squid, octopus, one chiton (the gumboot), and a pelecypod. Presence of an external calcareous shell is fair evidence that the occupant is a mollusk, but a few non-mollusks also build or inhabit such structures: brachiopods, polychete worms, barnacles, and hermit crabs (the shells of urchins are considered internal, since they are covered by a thin skin).

Major subdivisions include the Amphineura (chitons); Scaphopoda (tooth shells); Gastropoda (snails, abalones, nudibranchs, and sea hares); Pelecypoda (clams, mussels, oysters, scallops); and Cephalopoda (squid, octopus). Mollusks occur almost everywhere, ranging from a few isolated and unimportant individuals to massive aggregates profoundly influencing the character of a community.

CLASS AMPHINEURA–THE CHITONS

Chitons are sluggish, oval-shaped animals, tightly adhering to rocks through suction exerted by the foot. The protective shell is subdivided into 8 sections or valves, surrounded laterally by a tough strip of tissue, the girdle. Most chitons retire to dark crevices or the undersides of rocks during daytime.

Callistochiton palmulatus: dark brown; anterior valve with 11 ribs, posterior with 7 bifurcated ribs, length 3/8-inch (0.9cm). Central California to Baja California, low intertidal to 80 feet (24.5m).

Chaetopleura gemma: red, orange, olive, or yellow; valves sculptured centrally with longitudinal beaded rows, laterally with radiating beaded rows, girdle bearing short hairs; length 3/5-inch (1.5cm). Central California to Baja California, low intertidal to 100 feet (30.5m).

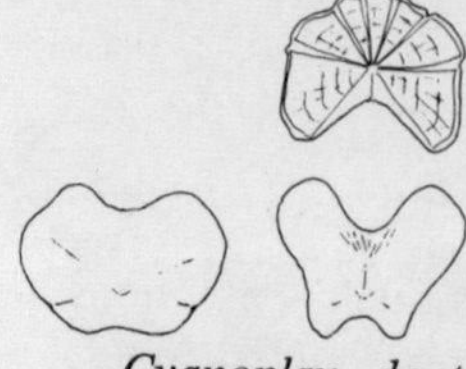

Cryptochiton stelleri (Gumboot): black or dark red-brown, girdle covers valves completely; length up to 12 inches (30.5cm). Northern and central California, low intertidal to 60 feet (18m).

Cyanoplax dentiens: gray, green, olive, or brown, darkly speckled; valves slightly beaked, anterior valve with 8 slits, minute papillae on girdle; ½-inch (1.3cm) long. Entire coast, intertidal to 40 feet (12m).

Cyanoplax hartwegi: olive green, sometimes lined or blotched; oval or almost round, 10-11 slits on anterior valve, girdle granular; up to 1 inch (2.5cm). Entire coast, low intertidal and shallow subtidal.

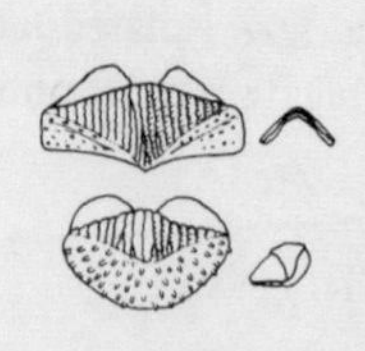

Ischnochiton (several species): girdle with fine scales, head and tail valves equal size, central part of tail valve raised. *I. fallax*: marbled rosy-orange, valves smooth except for central pitting; length 1 inch (2.5cm). Entire coast, low intertidal to 250 feet (46m). *I. mertensii* (illustrated): yellow, orange, red-brown, sometimes speckled white; valves strongly sculptured with ribs and nodules; 1½ inches (3.8cm) long. Entire coast, intertidal to 60 feet (18m). *I. regularis*: olive, gray, or blue; delicate longitudinal sculpturing on valves centrally, radial striations criss-crossing concentric ridges laterally; 2 inches (5.1cm) long. Northern and central California, shallow.

Katherina tunicata: white valves, two-thirds covered by smooth black girdle; up to 3 inches (7.6cm) long. Northern to southern California, low intertidal, occasionally shallow subtidal.

Mopalia ciliata (Hairy Chiton): color variable, usually green with dark or light splotching; girdle yellow to brown, cleft posteriorly, bearing profusion of curling brown hairs with white spines basally; 1½ inches (3.8cm) long. Entire coast, intertidal to 40 feet (12m).

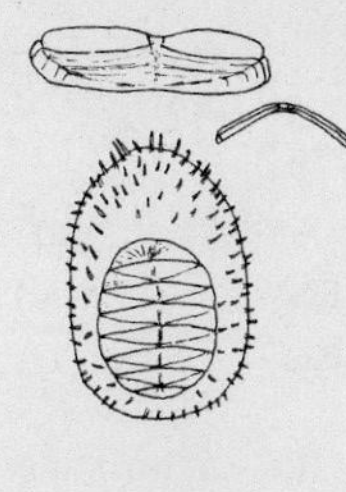

Placiphorella velata: dull red or olive-brown, yellow-red girdle, wide anteriorly with diamond-shaped scales and sparse stiff hairs; 2 inches (5.1cm) long. Entire coast, low intertidal to shallow subtidal.

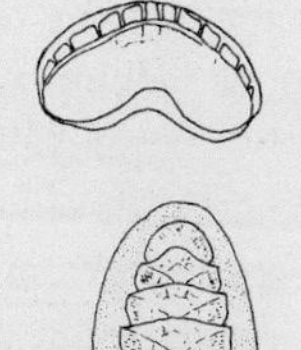

Stenoplax (several species): girdle scaled, tail valve larger than head. *S. heathiana*: drab green or gray, eroding to white; valves smooth, girdle scales large; up to 3 inches (7.6cm) long. Entire coast, shallow. *S. conspicuus* (illustrated): green and pinkish mottled valves; elongated girdle scales resemble stubby bristles; up to 6 inches (15.2cm). Southern California and Baja California, intertidal to 50 feet (15m).

Tonicella lineata: lavender, red, brown, or yellow background, crossed with zigzag lines of orange, red, green, and blue; girdle smooth; 1 inch (2.5cm) long. Entire coast, but not common south of Point Conception; low intertidal to 180 feet (55m). *T. marmorea* similar but darker and without pattern of lines.

CLASS SCAPHOPODA–THE TOOTH SHELLS

The scaphopods are a rather minor class of mollusks, resembling somewhat the long pointed tusks of walruses or elephants. The animals bury themselves in sediments and the white shells rarely exceed 1 inch (2.5cm) in length; hence they are fairly inconspicuous. In some areas, however, empty scaphopod shells accumulate in great numbers in depressions in the sand, and presumably the animals are an important component of the fauna in such places.

Cadulus: maximum diameter of shell occurs posterior to aperture. Entire coast, 35 feet (11m) or deeper.

Dentalium: maximum diameter of shell occurs at aperture. *D. neohexagonum* is hexagonal in cross-section. Central California to Baja California, 30 to 350 feet (9-107m).

CLASS GASTROPODA–SNAILS, ABALONES, NUDIBRANCHS, etc.

Most gastropods have a muscular, flexible, flat organ called the *foot* which is applied against the substrate for clinging or movement. The foot is present even in gastropods lacking an external shell (sea hares) or without any shell (nudibranchs). The only other common groups displaying a clinging foot-like organ are the chitons and a few sea cucumber species (*Psolus* and *Thyonepsolus*). To facilitate identification in the field, it is convenient to divide the gastropods into the following groups, based on shell characteristics:

I. Open aperture; shell dishlike or tentlike, spire reduced or absent
II. Elongate aperture, at least half shell length; spire reduced
III. Triangular-oval; spire angle 40° or more, base flat or round
IV. Spindle-shaped; spire angle 40° or more, base angular
V. Pointed; spire angle 30° or less
VI. Tubular
VII. Soft-bodied; shell internal
VIII. Soft-bodied; shell lacking (nudibranchs)

Spire angle refers to the angle formed at or above the apex of the shell by lines passing down along the edges of the coiled portion, excluding bumps, ridges, wings, spines, etc. In a few cases there is overlap among groups III, IV, and V, due primarily to individual variation in shell shape; in such cases we have listed the species in two groups.

I. OPEN APERTURE, SHELL DISHLIKE OR TENTLIKE

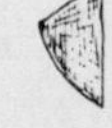
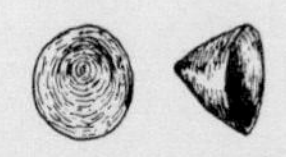

Acmaea (limpet; several species): *A. insessa* (illustrated, top): brown shell, up to 1 inch (2.5cm) long; on stipes of feather-boa kelp (*Egregia*). Entire coast, intertidal to 40 feet (12m). *A. mitra* (illustrated, middle): shell white, often pink due to coralline encrustation; thick, 1½ inches (3.8cm). Entire coast, intertidal to 150 feet (45m). *A. paleacea* (illustrated, bottom): shell brown with white top, elongate, ¼-inch (0.6cm); on surf grass (*Phyllospadix*). Entire coast, intertidal to 20 feet (6m).

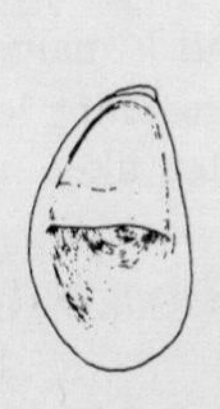

Crepidula (Slipper Shell): shell usually tan to dark brown, apex bent over toward margin. Inside, shell forms tiny shelf (*deck*) attached along both edges beneath apex. *C. adunca*: apex high and strongly recurved; ½-inch (1.3cm). Entire coast, intertidal to 80 feet (24m). *C. excavata*: shell whitish mottled with brown, apex curled beneath itself; 1 inch (2.5cm) long. Central California to Baja California, intertidal to 40 feet (12m). *C. nummaria*: shell white, usually covered by brown periostracum; flattened, deck thin, 1½ inches (3.8cm) long. Entire coast, intertidal to 60 feet (18m). *C. onyx* (illustrated): shell dark brown both inside and outside, except for white deck; 2 inches (5.1cm). Central California to Baja California, intertidal to 100 feet (30.5m).

Crepipatella lingulata (Wrinkled Slipper Shell): shell brown, wrinkled, circular, thin, flat, interior deck attached only along one side; ¾-inch (1.9cm). Entire coast, intertidal to 80 feet (24m).

Diodora (Keyhole Limpet): round or slightly oval aperture at apex (distinguishes from *Fissurella*, below). *D. aspersa*: shell gray-white with purple radiating bands, posterior edge of aperture much higher than anterior; up to 3 inches (7.6cm). Entire coast, intertidal to 100 feet (30.5m). *D. murina*: shell white with gray-black radiating rays, posterior edge of aperture slightly higher than anterior; 1 inch (2.5cm) long. Entire coast, low intertidal to 60 feet (18m).

Fissurella volcano (Keyhole Limpet): elongate aperture at apex 3 times longer than broad (distinguishes from *Diodora*, above); shell white or pink with red-purple rays, up to 1 inch (2.5cm); animal yellow with red striping. Entire coast, intertidal to 60 feet (18m).

Haliotis (Abalone): tiny spire with flattened whorls, final whorl is main part of shell; row of perforations along left margin. *H. assimilis* (Threaded Abalone); 4 to 6 raised apertures, moderately high spire, shell greenish, 5 inches (12.5cm) in diameter; body yellow with brown blotches. Southern California and Baja California, 10 to 100 (3-30.5m) or more feet. *H. corrugata* (Pink Abalone) (illustrated, top): 2 to 4 raised apertures, shell green to red-brown with even wavy corrugations, margin scalloped, 10 inches (25cm) though usually smaller; body mottled black and white. Southern California and Baja California, intertidal to 180 feet (55m). *H. cracherodii* (Black Abalone): 5 to 9 flush apertures (sometimes none), shell smoothish and black, 7 inches (17.8cm); body black. Entire coast, intertidal to 20 feet (6m). *H. fulgens* (Green

Abalone): 5 to 7 holes, shell brownish with low ridges, usually heavily encrusted, 10 inches (25cm); body blackish, mottled. Southern California and Baja California, intertidal to 80 feet (24m). *H. kamtschatkana* (Pinto Abalone): 3 to 6 holes, high spire, shell greenish brown, mottled, with ridges and lumps, 4 inches (10.2cm) common; body tan to greenish-brown. Northern and central California, 35 to 50 feet (10-15m). *H. rufescens* (Red Abalone) (illustrated, bottom): 3 or 4 holes, shell brick red with occasional white stripes, small ridges and elongated lumps, up to 10 inches (25cm); black body. Entire coast, intertidal to 500 feet (150m) in cold water. *H. sorenseni* (White Abalone): 3 to 5 holes, shell highly arched, reddish brown, irregular ridges, 5 to 8 inches (12.5-20.5cm) common; yellow body mottled with green to brown. Common around islands, southern California and Baja California, 15 to 150 feet (5-45.5m). *H. walallensis* (Flat Abalone): 4 to 8 holes, shell flattened, dark reddish with blue-green or white mottling, with few ridges, regular longitudinal striations, 3 to 5 inches (7.6-12.5cm) common; body mottled yellow and brown. Entire coast, intertidal to 90 feet (27m).

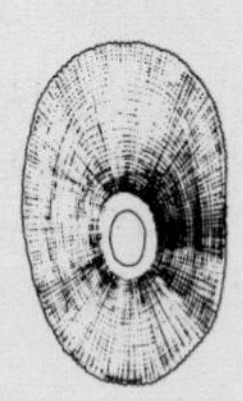

Megathura crenulata (Giant Keyhole Limpet): black or brown mantle conceals most of whitish shell; oval aperture at apex, fine radiating lines, 4 inches (10.2cm). Central California to Baja California, intertidal to 110 feet (34m).

II. ELONGATE APERTURE, SPIRE REDUCED

Bulla gouldiana (Bubble Shell): yellow to brown shell, mottled, thin; aperture extends entire length, pit at apex; 2 inches (5.1cm) long. Primarily in bays on mud or sand; southern California and Baja California, intertidal to 30 feet (9m).

Conus californicus (Cone Snail): tan to brown shell, conical at both ends, narrow aperture extends three-fourths of length; up to 1 inch (2.5cm) long. Central California to Baja California, intertidal to 150 feet (45.5m).

Erato (several species): pear-shaped shell with flat coiled spire; aperture extends three-fourths of length, with teeth in outer lip. *E. columbella*: shell olive-brown, ¼-inch (0.6cm). *E. vitellina*: shell red-brown, ½-inch (1.3cm). Central California to Baja California, intertidal to 300 feet (91.5m).

Haminoea (White Bubble Shell): shell pale yellow to green, thin, fragile, transparent; smooth, flaring outer lip lacks reinforcing ridge at apex (distinguishes from *Bulla*, above); animal yellow-brown. *H. vesicula*: aperture and body whorl equal size; ¾-inch (1.9cm) long. Entire coast. *H. virescens*: aperture larger than body whorl; ½-inch (1.3cm). On rocks and sand, bays and open coast; southern California and Baja California, intertidal to 60 feet (18m).

Marginella californica: shell white to brown, often banded, smooth lustrous surface; spine evident, outer lip smooth, inner lip has basal folding; 3/8-inch (1.0cm) long. Southern California and Baja California, intertidal and shallow subtidal.

Neosimnia inflexa (Egg Shell): shell purple, lavender, rose, or red, delicate; narrow, very elongate aperture, inner lip smooth and twisted posteriorly; ½-inch (1.3cm) long. Common on gorgonian corals (Sea Fans). Central California to Baja California, 60 feet (18m) or deeper.

Olivella (Olive Shell): shell approximate size and shape of large olive pit, glossy pearl-gray with white, brown, or purple shading; inner lip has twisting fold anteriorly. *O. baetica*: shell elongate, pointed spire, greatest diameter near base. *O. biplicata* (illustrated): spire relatively low, greatest diameter near posterior aperture. *O. pedroana* (*pycna*): spire prominent, diameter nearly uniform over entire aperture. On sediment bottoms, bays, and open coast; entire coast, intertidal to 40 feet (12m), sometimes deeper.

Pusula (*Trivia*) (Coffee Bean Shell): oval shell encircled by prominent ribs; aperture extends entire length. *P. californiana*: pink to red-brown shell, ribs joined dorsally in a white line or crease; 3/16-inch (0.5cm). Entire coast. *P. solandri* (illustrated): chocolate to dark pink shell, ribs meet dorsally in white groove with nodules marking junctions; ¾-inch (1.9cm). Southern California and Baja California, intertidal to 200 (61m) or more feet.

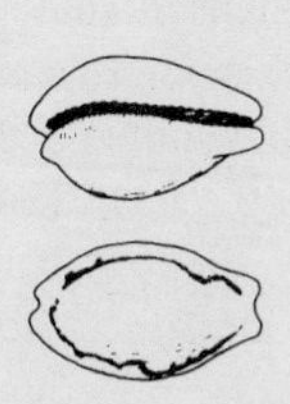

Zonaria (*Cypraea*) *spadicea* (Chestnut Cowrie): shell reddish-brown dorsally, shading to white ventrally, smooth and shiny, oval; aperture as long as shell, both lips with teeth; 2 inches (5.1cm) long. Central California to Baja California, low intertidal to 150 feet (45.5m).

III. TRIANGULAR-OVAL SHELL

Acteon punctocoelata (Barrel Shell): white shell with 2 spiraling dark bands, oblong, 4 to 5 whorls; ¾-inch (1.9cm). Entire coast, intertidal to 60 feet (18m).

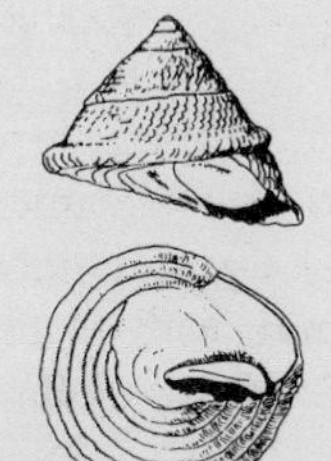

Astraea (Wavy Top Snail): shell margin undulated, calcareous operculum, base relatively flat with deep spiral grooves. *A. gibberosa* (*inequalis*): red shell, 3 inches (7.6cm) in diameter. Entire coast, cold water, intertidal to 350 feet (107m).

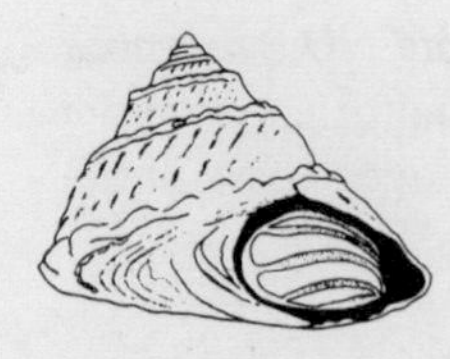

A. undosa: dull yellow-brown shell, 6 inches (15cm) in diameter. Southern California and Baja California, intertidal to 80 feet (24m).

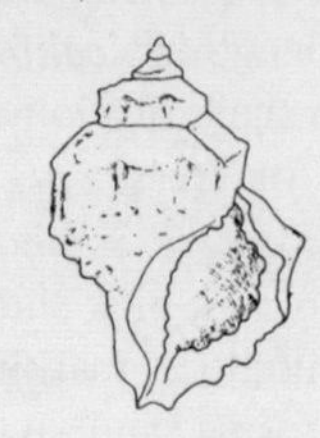

Bursa californica (Frog Snail): yellow-brown shell, white inside, 2 prominent longitudinal ridges with nodules giving lumpy appearance; notches at both ends of aperture; 3½ inches (8.9cm). Central California to Baja California, 35 to 350 feet (11-107m).

Calliostoma (Top Shell): conical outline, numerous smooth or finely beaded spiral ribs, lacking umbilicus; diameter to 1 inch (2.5cm). *C. annulatum*: alternating yellow and purple spiral banding. Entire coast, intertidal to 80 feet (24m). *C. canaliculatum*: ivory shell with thin dark spiral lines, sutures indistinct. Northern to southern California, intertidal to 90 feet (27m). *C. costatum* (*ligatum*): intertidal to 40 feet (12m). *C. gloriosum* (illustrated); salmon pink to yellow-brown shell with slanting dark purple mottling. Central and southern California, shallow subtidal to 100 feet (30.5m). *C. supergranosum*: chestnut brown shell with white spotting; ½-inch (1.3cm) in diameter. Central California to Baja California, intertidal to 60 feet (18m). *C. tricolor*: yellow-brown shell, 3 ridges per whorl have alternating white and purple spots. Northern California to Baja California, 50 to 200 feet (15-60m).

Lacuna (Chink Shell): thin delicate shell, globose outline, smooth round whorls, inner lip bordered by a groove; small, about 1/8-inch (0.3cm) long. *L. unifasciata*: early whorls pinkish, older whorls tan to brown; dotted ridge on bottom whorl. Animals common on seaweeds; entire coast, intertidal to 60 feet (18m).

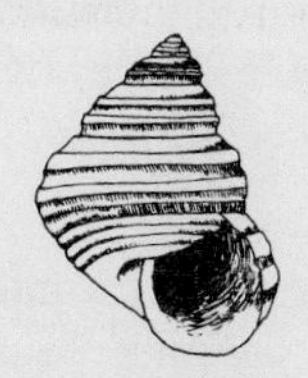

Margarites: smooth lustrous surface, rounded whorls with deep sutures, umbilicus in base; uniform coloring. *M. lirulata*: purple exterior, iridescent pearly interior; ¼-inch (0.6cm) in diameter. Entire coast, intertidal to shallow subtidal.

Nassarius (*Alectrion*) (Dog Whelk): pointed spire, prominent sculpture, oval aperture; short anterior canal separated from basal whorl by furrow joining prominent inner lip. *N. cooperi*: white, yellow, or brown shell with prominent longitudinal ribs, subdued spiral ridges; length ¾-inch (1.9cm). *N. fossatus*: white, yellow, orange, or brown shell; spiral ridges and longitudinal ribs form conspicuous nodes at intersection; 2 inches (5.1cm). *N. mendicus*: yellowish to light brown shell, ridges and ribs about equal in prominence; ¾-inch (1.9cm). *N. perpinguis* (illustrated): yellow-white to brown shell, occasionally banded; beaded spiral ridges prominent. Commonly on sediment bottoms, bays or open coast; entire coast, intertidal to 60 feet (18m).

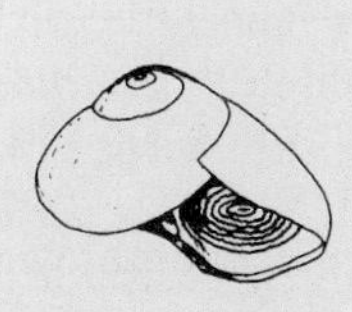

Norrisia norrisii (Smooth Turban): smooth, thick shell, red exterior, black at center of base, pearly interior; umbilicus present; 2 inches (5.1cm). Central California to Baja California, intertidal to 100 feet (30.5m).

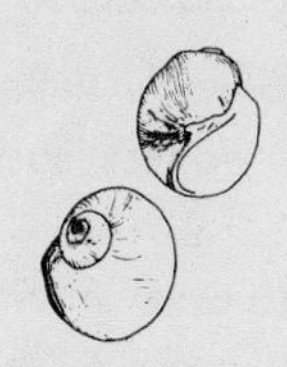

Polinices (Moon Snail): globose outline, small spire, brown or white enamel of inner lip partially overlapping umbilicus. *P. draconis*: brown, gray, or white shell, diameter slightly greater than height; 3 inches (7.6cm) diameter. *P. lewisii* (illustrated): brown, gray, or yellowish-white shell, diameter slightly less than height; 5 inches (12.7cm) in diameter. *P. recluziana*: white, bluish gray, or brown shell, enamel frequently obscures umbilicus, diameter exceeds height; 3 inches (7.6cm) in diameter. Entire coast, intertidal to 150 feet (46m).

Pteropurpura (Pterynotus) trialatus (Three-Wing Murex): chestnut to dark brown shell, 3 prominent thin convoluted longitudinal plates with finely striated faces anteriorly, with or without basal tubercle. Entire coast, shallow subtidal to 100 feet (30.5m) or more.

Tegula (Turban Snail): conical outline, often slightly inflated; flattish base, horn-like operculum, pearly interior, teeth on inner lip. *T. aureotincta*: gray to black shell, bright yellow patch basally around umbilicus; 1 inch (2.5cm) long. Southern California and Baja California, low intertidal to 40 feet (12m). *T. brunnea* (illustrated): tan, orange, or red-brown shell, whorls slightly rounded, umbilicus closed; 1 inch (2.5cm). Entire coast, low intertidal and shallow subtidal. *T. montereyi*: olive to brown shell, whorls and base flat, giving conical shape; tooth on inner lip, umbilicus funnel-shaped; 1 inch (2.5cm). Central California to Baja California, intertidal to 50 feet (15m). *T. pulligo*: brown-purple shell, shaped like *T. montereyi* but lacks tooth on inner lip and umbilicus partly covered by enamel; 1 inch (2.5cm). Northern to southern California, intertidal to 50 feet (15m). *T. regina*: tan to black shell with bright gold patch basally, base and whorls flat, giving conical shape; slanting axial ridges, scalloped margin; 2 inches (5.1cm). Southern California and Baja California, 50 to more than 100 feet (15-30.5m).

Tricolia compta (Pheasant Shell): translucent gray-green

periostracum, zigzag and spiral white, brown, or red spots and lines beneath; rounded whorls, prominent spire; up to ½-inch (1.3cm). Bays and open coast, entire coast, intertidal to 80 feet (24m).

IV. SPINDLE-SHAPED SHELL

Amphissa versicolor: yellow, gray, red, or black shell, sometimes striped or mottled; spiral ribbing on entire surface, longitudinal ridges stop at level of aperture; ½-inch (1.3cm). Entire coast, intertidal to 80 feet (24m).

Bursa californica: canal and sculpturing sometimes confuse appearance, described in group III (above).

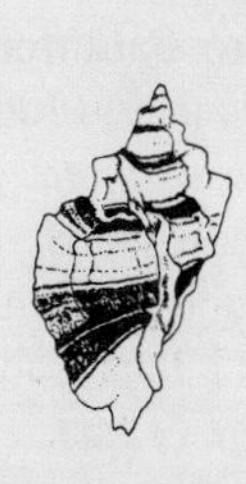

Ceratostoma (*Pterorytis, Purpura*) (Hornmouth): 3 conspicuous longitudinal plates (*varices*), ridges in between, tooth on lower outer lip, canal closed. *C. foliata* (illustrated): dull white shell with brown striping, varices made of overlapping layers; up to 3 inches (7.6cm) long. Entire coast, intertidal to 200 feet (61m). *C. nuttallii*: drab yellow to brown shell, varices not much larger than axial ridges, 3 teeth within outer lip; 2 inches (5.1cm) long. Southern California and Baja California, intertidal to 40 feet (12m).

Fusinus (Spindle Shell): light tan or yellow shell with dark spiral lines, deep sutures, moderate longitudinal ribs; aperture base narrows to long channel with no complex folding at bottom. *F. kobelti*: whitish, 2 inches (5.1cm). *F. luteopictus*: dark brown between ribs, ¾-inch (1.9cm). Central California to Baja California, intertidal to more than 100 feet (30.5m).

Jaton festivus (Festive Rock-Shell): adult shell drab white, juvenile shell scarlet; 3 prominent thin ridges terminating above in recurved points, deep sutures, fine spiral ridges; 1½ inches

(3.8cm). Southern California and Baja California, intertidal to 450 feet (137m).

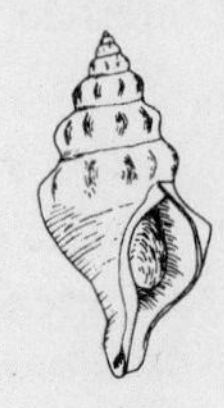

Kelletia kelletii (Kellet's Whelk): thick, whitish shell with sinuous suture, prominent knobs, long curving canal; 6 inches (15cm); orange body. Southern California and Baja California, intertidal to 120 feet (37m).

Maxwellia: thick vertical ridges, long anterior canal beneath aperture; 1½ inches (3.8cm) long. *M. gemma* (illustrated): white shell with brown striping, square pits in spire. *M. santarosana*: brownish white shell, low spire, tips of vertical ridges recurved. Southern California and Baja California, intertidal to 200 feet (61m) or more.

Megasurcula carpenterianus (Tower Shell): spire angle sometimes intermediate between groups IV and V; described under group V (below).

Mitra (Miter Shell): dark brown or black shell, elongate, sutures not prominent, fine surface sculpture, 3 conspicuous folds on inner lip; animal pure white. *M. catalinae* and *M. idae* (illustrated) similar except latter bigger–2 inches (5.1cm) vs. ¾-inch (1.9cm)–and with thicker shell. Entire coast, cold water, shallow subtidal to 200 feet (61m).

Mitrella (several species): spire angle intermediate between groups IV and V; described in group V (below).

Mitromorpha (Miter-form Shell): aperture narrow, about half shell length, short anterior canal; ¼-inch (0.6cm) long. *M. aspersa*: light orange to brown shell, beaded sculpture.

M. filosa: brownish orange to black shell, spiral striations, sutures inconspicuous. Central California to Baja California, low intertidal to 40 feet (12m).

Ocenebra (Dwarf Triton): oval to round aperture, inner lip folding over anterior canal paralleling conspicuous fold (*fasciole*), outer lip scalloped due to spiral ridges. *O. circumtexta* (illustrated): ivory to gray shell with brown spiral spotting, outer lip with teeth inside, prominent spiral grooves, low longitudinal ridges; 1 inch (2.5cm) long. Entire coast. *O. interfossa*: yellow, gray, or brown shell, axial and spiral ridges both prominent; ¾-inch (1.9cm). Entire coast. *O. paulsoni*: gray to brown shell crossed with fine white spiraling, knobby longitudinal ridges, teeth within outer lip; 1½ inches (3.8cm). Southern California and Baja California, intertidal to 80 feet (24m).

Pteropurpura (*Pterynotus*) *trialatus*: sculpturing confuses classification; described in group III (above).

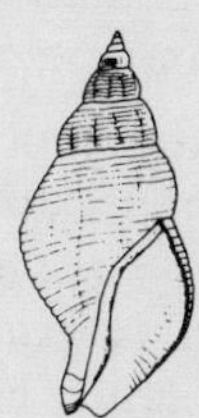

Searlesia dira: gray covering, red-brown beneath; spiral and axial ribbing on spine, spiral lines on base and on interior of outer lip, aperture oval with prominent canal. Northern and central California, intertidal and shallow subtidal.

Thais (*Nucella*) *lamellosa* (Dogwinkle): whitish, yellow, or brown shell, plain or banded, with prominent spire; spiral ridges throughout, forming convolutions in outer lip, conspicuous thin longitudinal scales or plates (unworn specimens); 3 inches (7.6cm) long, often aggregating. Northern and central California, intertidal to 20 feet (6m); common northward.

Urosalpinx cinerea (Oyster Drill): yellow to gray shell, reddish interior, subdued longitudinal ribs and spiral ridges, outer lip smooth and uniformly curving, canal straight; 1 inch

(2.5cm) long. Introduced with transplanted East Coast oysters in bays. Central and southern California, intertidal to 25 feet (8m).

V. POINTED SHELL

Alvania acutilirata (Risso-shell): brown shell, smooth round aperture, spiral ribbing with fine longitudinal lines; minute, up to 1/8-inch (0.3cm). Southern California and Baja California, 40 to 100 feet (12-30.5m).

Balcis rutila: white shell, sutures poorly developed, commensal on *Astropecten*; ½-inch (1.3cm) long. Central California to Baja California, intertidal to 100 feet (30.5m).

Bittium (several species): white to brown shell, small nodules regularly arranged to form spirals and ridges, aperture narrows anteriorly; generally small, ¼- to ½-inch (0.6-1.3cm) long. Entire coast, intertidal to 1000 feet (305m).

Burchia redondoensis: dark brown shell, gentle longitudinal ridging, spiral swelling below suture resembles second suture; 2 inches (5.1cm) long. On sediment bottoms. Southern California and Baja California, 90 to 150 feet (27-45m).

Epitonium (Wentle-trap Shell; several species): white or whitish round whorls, conspicuous smooth longitudinal ribbing extending over entire basal whorl, circular aperture; ½-inch to 1 inch (1.3-2.5cm) long. Often associated with anemones. Entire coast, throughout scuba zone. *E. sawinae* illustrated.

Megasurcula carpenterianus (Tower Shell): red-brown shell becoming lighter basally, with fine striations and ridges; aperture narrow, about one-third shell length, notched near suture; 3 inches (7.6cm) long; body orange. On sediment bottom, central and southern California, 60 feet (18m) and deeper.

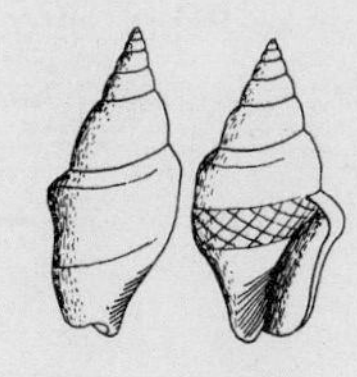

Mitrella (*Columbella*) (Dove Snail): spiral ribbing basally, upper part of spire smooth, glossy; oval aperture with short canal anteriorly. *M. aurantiaca*: translucent orange, pink, or brown shell. *M. carinata* (illustrated): brown or red shell, sometimes with white lines or spots; prominent ridge near suture of basal whorl. *M. gouldii*: similar coloring but lacking ridge. *M. tuberosa*: translucent tan shell flecked with brown or white; outer lip thickened with row of nodules inside. Lengths ¼- to ½-inch (0.6-1.3cm). On seaweeds, entire coast, intertidal to 110 feet (34m).

Ophiodermella (*Clathrodrillia, Moniliopsis*) *incisa* (Drill): ashen, brown, or white shell, reddish spiral grooves, sinuous longitudinal ridges, aperture notched posteriorly; 1 inch (2.5cm) long. Entire coast, 30 to more than 100 feet (9-30.5m).

Seila monteryensis: yellowish to red shell, flat whorls causing uniform taper, sutures concealed by strong spiral grooves; length ½-inch (1.3cm). Central California to Baja California, intertidal to 400 feet (122m).

Terebra pedroana (Auger Shell): gray or brown shell, flat whorls with beaded ridge following suture, short recurved canal; 1¼ inches (3.2cm) long. On sediment bottoms, southern California and Baja California, 30 to 100 feet (9-30.5m).

Turbonilla (several species): numerous, uniform longitudinal ridging. *T. choclata*: shiny golden brown shell, pointed apex; length ½-inch (1.3cm). Central and southern California, intertidal to 150 feet (46m). *T. kelseyi*: light colored, semitransparent shell with truncated apex; ¼-inch (0.6cm). Southern California and Baja California, intertidal to 200 feet (61m). *T. laminata*: yellow shell, axial ridges interrupted by uniform spiral grooves giving beaded appearance; ¼-inch (0.6cm). Southern California and Baja California, intertidal to 150 feet (46m). *T. tenuicola*: white to brown shell, minutely reticulated; ¼-inch (0.6cm). Central California to Baja California, intertidal to 50 feet (15m). *Turbonillas* often associate with polychete worms and bivalve mollusks.

VI. TUBULAR SHELL

Caecum californicum: white, slightly curved shell of 30 to 40 prominent rings; 1/8-inch (0.3cm) long. Central California to Baja California, intertidal to 40 feet (12m).

Dendropoma (Spiroglyphis) lituella (Crooked Tube Snail): drab white, tube attached to rocks, recurving in rough spiral, on shells and rocks; tube 1/16-inch (0.15cm) in diameter. Entire coast, intertidal to 90 feet (27m).

Micranellum crebricinctum (Tube Shell): whitish, slightly curved tube, smooth except for fine encircling rings; ¼-inch (0.6cm) long. Resembles *Caecum* (above), but shell smoother. Central California to Baja California, intertidal to 100 feet (30.5m).

Petaloconchus (Macrophragma) montereyensis (Tube Snail): brownish, attached shell irregularly uncoiling with growth, transverse ridging, operculum present; ¼-inch (0.6cm) long. Central California to Baja California, intertidal to 100 feet (30.5m).

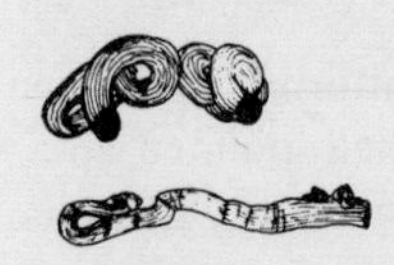

Serpulorbis (Aletes) squamigerus (Tube Snail): dingy white or gray, minutely scaled tubes, transverse circular ridges, operculum lacking; ¼-inch (0.6cm) in diameter, forms large convoluted masses. Entire coast, intertidal to 60 feet (18m).

VII. SOFT-BODIED, SHELL INTERNAL

Aglaja diomeda: reddish-brown to bluish-black, lateral folds not covering dorsal surface, dorsal groove defines head; ¾-inch (1.9cm) long. Usually just barely buried in sediment; bays and open coast, entire coast, intertidal to 100 feet (30.5m).

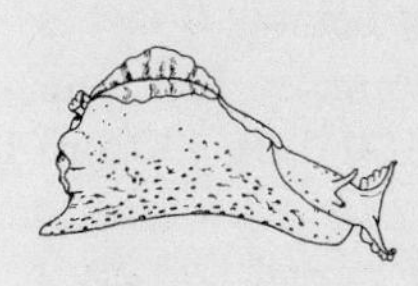

Aplysia (Sea Hare): 4 protuberances on head, 2 prominent dorsal folds. *A. californica* (illustrated): light brown, conspicuous mottling; 8 inches (20.3cm). Central California to Baja California, intertidal to 60 feet (18m). *A. vaccaria*: black, slightly mottled; 30 inches (76.2cm). Southern California and Baja California, intertidal to 40 feet (12m).

Chelidonura (*Navanax*) *inermis* (Striped Sea Slug): brown background with stripes and dots of white, yellow, and blue; prominent lateral folds meeting dorsally. Bays and open coast, central California to Baja California, intertidal to 40 feet (12m).

Petalifera (*Phyllaplysia*) *taylori*: green with black longitudinal stripes or rows of dots, pointed posteriorly; 2 inches (5.1cm) long. On *Zostera* (eel grass). Northern to southern California, low intertidal to 20 feet (6m).

VIII. SOFT-BODIED, SHELL LACKING

Sixteen common nudibranchs are shown in color as an aid to identification (Plates 5 and 6). Consequently, comments below will emphasize geographic and depth ranges, lengths, and special habitat features. If color varies, a range of hues is given.

Aegires albopunctatus: ½-inch (1.3cm). Entire coast, intertidal to 100 feet (30.5m).

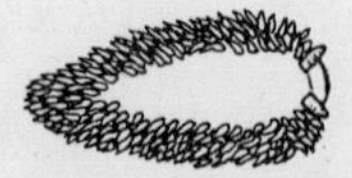

Aeolidiella papillosa: 2 inches (5.1cm). Northern and central California, intertidal to 10 feet (3m).

Aldisa sanguinea: ½-inch (1.3cm). Central California to Baja California, intertidal to 80 feet (24m).

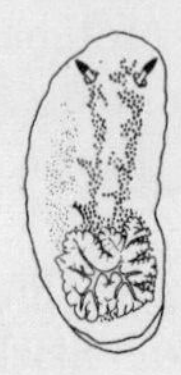

Anisodoris nobilis: black splotches variable and lie between tubercles (distinguishes from *Archidoris* below); 8 inches (20.3cm). Central California to Baja California, intertidal to 100 feet (30.5m) (Plate 6h).

Archidoris montereyensis: whitish to light yellow, black spots variable and cover tubercles (distinguishes from *Anisodoris*, above); 2 inches (5.1cm). On sedimentary bottoms, entire coast, shallow subtidal to 300 feet (91m) (Plate 6a).

Austrodoris odhneri: white, sometimes with yellow tinges; up to 4 inches (10cm). Entire coast, shallow subtidal to 60 feet (18m) (Plate 5d).

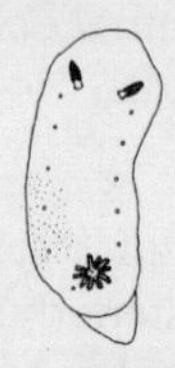

Cadlina flavomaculata: ¾-inch (1.8cm). Entire coast, intertidal to 100 feet (30.5m).

Cadlina limbaughi: 1 inch (2.5cm). Southern California and Baja California, 30 feet (9m) to deeper than 100 feet (30.5m) (Plate 6c).

Cadlina marginata (*luteomarginata*): 1½ inches (3.8cm). Entire coast, intertidal to 110 feet (34m) (Plate 6e).

Chioraera (*Melibe*) *leonina*: 1 inch (2.5cm). On kelp and large seaweeds well out from shore. Entire coast, surface to 35 feet (11m) (Plate 5c).

Corambe pacifica: ½-inch (1.3cm). On kelp and large offshore seaweeds. Northern to southern California, surface to 20 feet (6m).

Dendrodoris (*Doriopsis*) *albopunctata*: yellow to orange feathery antennae; 2 inches (5.1cm). *D. fulva*: similar except for conical, pointed, orange-brown antennae. Entire coast, intertidal to 150 feet (46m) (Plate 6f).

Dendronotus frondosus: colorless or yellow, orange, or red; uniform, banded, or mottled; 2 inches (5.1cm). Northern to southern California, cold water, intertidal to 1200 feet (366m) (Plate 6g).

Diaulula sandiegensis: white to tan, numbers of black spots vary; 3 inches (7.6cm). Entire coast, intertidal to 100 feet (30.5m).

Duvaucelia (*Tritonia*) *festiva*: colorless to greenish; 1½ inches (3.8cm). Entire coast, intertidal to 120 feet (37m) (Plate 5a).

Flabellinopsis (*Flabellina*) *iodinea*: 3 inches (7.6cm). Entire coast, intertidal to 100 feet (30.5m) (Plate 6d).

Glossodoris (*Chromodoris*) *californiensis*: 2 inches (5.1cm). Central California to Baja California, intertidal to 60 feet (18m) (Plate 5g).

Glossodoris (*Chromodoris*) *macfarlandi*: ½-inch (1.3cm). Central to southern California, intertidal to 80 feet (24m).

Glossodoris (*Chromodoris*) *porterae*: ½-inch (1.3cm). Central to southern California, intertidal to 80 feet (24m) (Plate 5e).

Hermissenda crassicornis: 3 inches (7.6cm). Entire coast, intertidal to 100 feet (30.5m) (Plate 6b).

Hopkinsia rosacea: 1 inch (2.5cm). Central California to Baja California, intertidal to shallow subtidal (Plate 2a).

Laila cockerelli: ¾-inch (1.8cm). Northern to southern California, intertidal to shallow subtidal (Plate 5h).

Phidiana nigra: 2 inches (5.1cm). Often accompanies *Hermissenda* (above). Central to southern California, intertidal to 60 feet (18m) (Plate 5b).

Polycera atra: 1 inch (2.5cm). Central to southern California, intertidal to shallow subtidal.

Rostanga pulchra: pink to red; ¾-inch (1.8cm). Often camouflaged against red sponges. Entire coast, intertidal to 80 feet (24m).

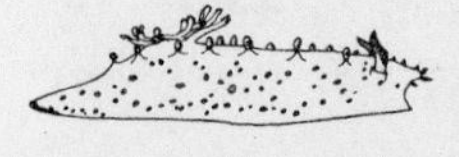

Triopha carpenteri: 1 inch (2.5cm). Northern to southern California, intertidal to 80 feet (24m) (Plate 5f).

Triopha grandis: 1 inch (2.5cm). Northern to southern California, intertidal to shallow subtidal.

CLASS PELECYPODA–BIVALVE MOLLUSKS

The Pelecypods (Lamellibranchs) possess two shells (*valves*) hinged together and joined by a flexible ligament. The valves may or may not enclose the entire soft body (one species, *Chlamydochonca orcutti*, has internal shells). In some species the valves are mirror images, while in others one valve is different from its mate. Existence of two valves identifies an animal as a Pelecypod–with one exception. Brachiopods are also bivalved, but the only local common species at depths frequented by divers is *Terebratalia transversa* (see section on Brachiopods).

Pelecypods are ubiquitous, occurring in bays, harbors, and open coasts and lying shallow or deep and in or on rock, sand, or mud. The various species are almost always restricted in their environments and living habits, making such characteristics useful in identification, particularly in the field. We have used these characteristics to divide the Pelecypods into small convenient groups, but some caution is necessary when using this system: our classification refers to the living animals; after death the shells are often moved to areas quite different from the normal environment. Do not assume that an empty shell found on sand, for example, means that the animal was a sediment-burrowing form. Occasionally strong currents or waves will temporarily expose burrowing forms; such animals will quickly try to bury themselves again if given a chance.

The subclassifications we use here do not in any way reflect family or other evolutionary relationships. They are merely indications of common living habits or habitats.

I. Forms attached to rocks, pilings, etc.
II. Forms boring in rock, shell, etc.
III. Forms burrowing in sand and mud
IV. Forms nestling in crevices, etc., or free-living

I. ATTACHED FORMS

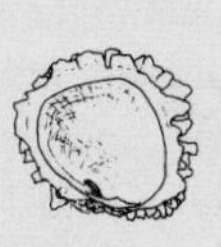

Chama pellucida (Rock Clam): pinkish-white to drab; uneroded shells have flattened spines, circular outline; 1 inch (2.5cm) across. Distinguish from *Pseudochama* (below) by counterclockwise growth in shell (viewed from above). Entire coast, intertidal to 150 feet (46m).

Glans (*Cardita*) *carpenteri* (*minuscula, subquadrata*): drab white to brown shell, quadrate, beaks anterior, conspicuous radial ribs; ¾-inch (1.8cm). In crevices and holes; entire coast, intertidal to 300 feet (91m).

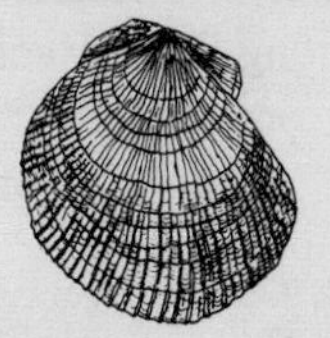

Hinnites multirugosus (*giganteus*) (Rock Scallop): dull tan or brown shell, often encrusted, purple stain interiorly near hinge; circular scallop appearance often altered by impinging rocks, etc.; 3 to 5 inches (7.6-12.7cm) in diameter. Free-living when young. Entire coast, intertidal to 180 feet (55m).

Kellia (several species): white shell, green to light brown glossy periostracum, globose, siphons point in opposite directions; retractable mantle covers shell. *K. laperousi*: oval outline; 1 inch (2.5cm). *K. suborbicularis*: subcircular outline; ½-inch (1.3cm). Entire coast, intertidal to 200 feet (61m).

Leptopecten (*Pecten*) *latiauratus* (Kelp Scallop): shell yellow to brown with white mottling, thin, semi-transparent, fragile; 1 inch (2.5cm) in diameter. Attached to seaweeds or free-living. Central California to Baja California, surface to 120 feet (37m).

Modiolus (*Volsella*) (Horse Mussel): yellow to brownish shell, one edge straight, the other with a curved flaring lip; often sparse to dense hairs posteriorly, anterior edge projcts beyond beaks. Typically in bays, attached or nestling, shallow. *M. capax* (illustrated): densely hairy, shell reddish beneath brown periostracum, faint ribbing; 3 to 4 inches (7.6-10.2cm). Southern California and Baja California. *M. demissus*: lacks hairs, prominent radial ribbing. Central California. *M. modiolus*: hairy, light ribbing; up to 9 inches (22.9cm). Northern to southern California.

Mytilus californianus (Californian Mussel): black elongate shell tapering to a point at umbones, concentric ridges; up to 10 inches (25cm) long. Entire coast, usually intertidal but occasionally torn loose and found in shallow depths. *M. edulis* similar but smoothish shell.

Ostrea lurida (Oyster): drab whitish shell, circular to elongate oval, upper valve lightly ribbed or with wavy crenulations, single muscle scar within; 3 inches (7.6cm). Entire coast, intertidal to 120 feet (37m).

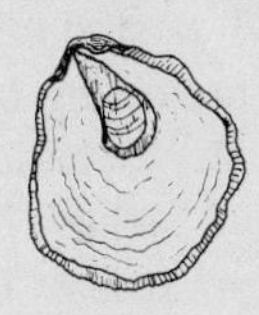

Pododesmus macroschisma (Jingle): drab tan shell, roughly circular, upper valve flat but roughened, lower valve with hole near hinge; 4 inches (10.2cm). Entire coast, low intertidal to 200 feet (61m).

Pseudochama exogyra (Rock Clam): drab whitish shell, sparse flattened spines, circular to oval; 2 inches (5.1cm) across. Distinguish from *Chama* (above) by clockwise growth in upper valve. Entire coast, intertidal to more than 60 feet (18m).

II. ROCK-BORING FORMS

Adula (*Botula*) *falcata* (Pea-pod Shell): brown, elongate, thin shell with fine transverse ridges crossing long growth lines, attaches within hole by a byssus fiber; 4 inches (10.2cm) long. Entire coast, intertidal to 40 feet (12m).

Chaceia (*Pholadidea*) *ovoidea* (Oval Piddock): dull white shell partly covered with brown rough periostracum, oval outline, beaks partly covered by recurving edge of shell. Entire coast, intertidal to 150 feet (46m).

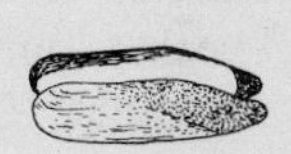

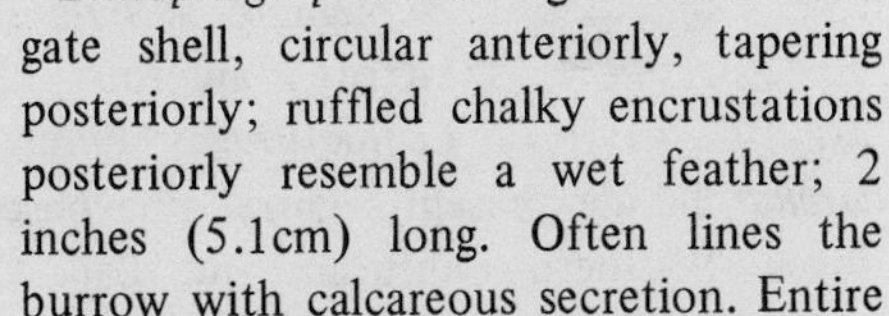

Lithophaga plumula: light brown elongate shell, circular anteriorly, tapering posteriorly; ruffled chalky encrustations posteriorly resemble a wet feather; 2 inches (5.1cm) long. Often lines the burrow with calcareous secretion. Entire coast, intertidal to 250 feet (76m).

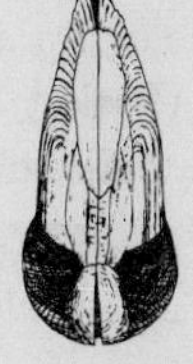

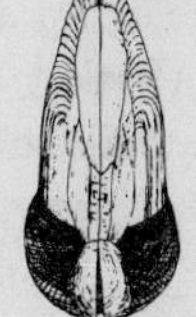

Parapholas (*Pholadidea*) *californica* (California Piddock): whitish shell partially covered by brown scaly periostracum along tapering posterior, beaks totally covered by recurving edge of shell; up to 5 inches (12.7cm). Entire coast, intertidal to 150 feet (46m).

Penitella (*Pholadidea*) *penita* (Common Piddock): whitish shell posteriorly covered by periostracum projecting beyond edge of shell, almost hemispherical anteriorly, beaks covered; 3 inches (7.6cm). Entire coast, intertidal to at least 60 feet (18m).

III. SEDIMENT-BURROWING FORMS

Acila (*Nucula*) *castrensis* (Nut Clam): trigonal shell, convex valves, exterior with intersecting ridges forming inverted V's; ½-inch (1.3cm). Entire coast, 30 to 4000 feet (9-1220m).

Amiantis callosa: shiny white oval shell, conspicuous concentric ridges, flaring anteriorly; up to 3 inches (7.6cm). Southern California and Baja California, intertidal to 60 feet (18m).

Chione (Hardshell Cockle): thick whitish to drab yellow shell, trigonal; concentric ridges crisscrossing radial ribs, beveled long smooth area behind ligament; 2 inches (5.1cm) long. *C. californiensis* (*succincta*) (illustrated): ribs evident throughout, *lunule* (heart-shaped sculpturing anterior to beaks) present. *C. fluctifraga*: lunule indistinct. *C. undatella*: lunule present; ventral ribs obscured by concentric ridges. Southern California and Baja California, common intertidally, sparsely out to 150 feet (46m).

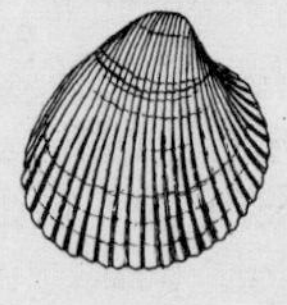

Clinocardium nuttalli (Basket Cockle): drab yellow-brown shell, radial ribs usually creased by transverse growth lines, beaks slant anteriorly; up to 6 inches

(1.9m). Northern to southern California, intertidal to at least 50 feet (15m).

Cryptomya californica (Softshell Clam): shell white, nearly smooth, grayish periostracum; right valve slightly deeper, crowding over left beak; 1¼ inches (3.2cm) long. Entire coast, intertidal to 300 feet (91m).

Diplodonta orbella: whitish shell, circular outline, almost resemble marbles; interior margin slightly scalloped; 1 inch (2.5cm). Often builds nest of detritus. Entire coast, intertidal and shallow subtidal.

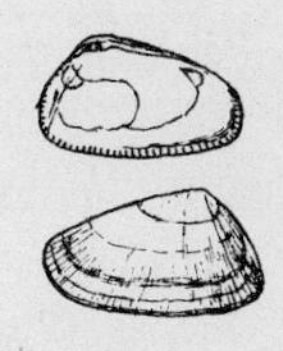

Donax (Bean Clam): shiny white, yellow, orange, gray, or purple shell, uniform or striped; oblong outline, inner margin crenulate; 1 inch (2.5cm). *D. californica*: angle at beaks much more than 90°. *D. gouldii* (illustrated): angle at beaks approximately 90°. Common, southern California and Baja California, intertidal and shallow subtidal, out to 90 feet (27.5m).

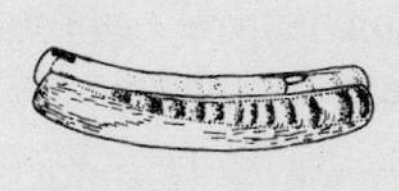

Ensis myrae (Razor Clam): purplish-white shell, brown periostracum, beak anterior, gently curving dorsal margin (distinguishes from *Solen* below); 2 inches (5.1cm). Central California to Baja California, intertidal to 150 feet (46m).

Gari (*Psammobia*) *californica* (Sunset Clam): cream-colored oval shell with faint radiating reddish-purple rays, thin wrinkled brown periostracum, concentric growth lines; up to 4 inches (10.2cm). Entire coast, intertidal to 150 feet (46m).

Laevicardium substriatum (Eggshell Cockle): smoothish tan shell with red-brown striping or mottling, fine radiating lines; 1 inch (2.5cm). Southern California and Baja California, intertidal and shallow subtidal.

Macoma: whitish to purple shell, often with gray or brown periostracum; thin, oval, smoothish, with low concentric ridges. *M. nasuta* (Bent-nosed Clam): bends to right side posteriorly when viewed from above; 2 inches (5.1cm) long. *M. secta* (White Sand Clam) (illustrated): left valve flatter than right; 4 inches (10.2cm). Other species occur. Entire coast, intertidal to about 25 feet (8m).

Mactra (several species): ligament divided into external and internal portions separated by thin plate of shell. *M. californica*: white shell, brown periostracum, trigonal and roughly symmetrical; concentric undulations near beaks; 2 inches (5.1cm). Entire coast, low intertidal to shallow subtidal in bays.

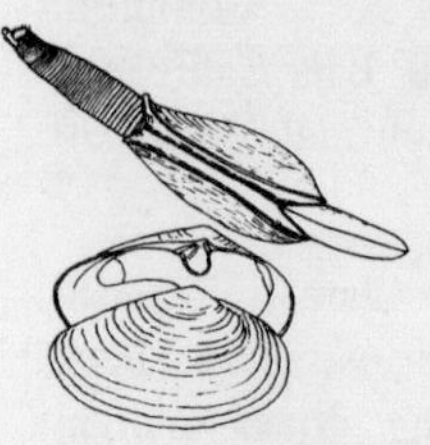

Mya arenaria (Softshell Clam): whitish oval shell, gray periostracum, beaks central, low concentric ridges; hinge has conspicuous tooth in one valve, inserting in socket of other valve; up to 5 inches (12.7cm). Northern and central California, intertidal and shallow subtidal.

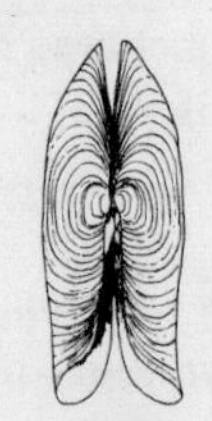

Panopea (*Panope*) *generosa* (Geoduck Clam): drab white shell, thin yellow periostracum, conspicuous concentric wavy ridging, quadrate; up to 9 inches (23cm). Entire coast, intertidal to 60 feet (18m).

Saxidomus (Washington Clam): gray to rust-colored shell, oblong, conspicuous concentric ridging; 3 to 4 inches (7.6-10.2cm). *S. gigantea*: shell interior pure white. Northern and central California. *S. nuttalli*: interior white, with purple markings posteriorly. Entire coast, intertidal to 120 feet (37m).

Siliqua (Jacknife Clam): shell smooth and glossy, fragile; beak central, interior reinforced centrally by rib. *S. lucida*: purplish translucent shell, periostracum lacking, interior rib vertical; 1½ inches (3.8cm). Central California to Baja California. *S. patula*: whitish-purple opaque shell, yellow-olive periostracum, interior rib oblique; 5 inches (12.7cm). On exposed beaches, northern and central California, intertidal to 200 feet (61m).

Solen (Razor Clam): fragile whitish shell, periostracum glossy olivine to yellow-brown, beak anterior (distinguishes from *Tagelus*, below); shell straight or only slightly curving (distinguishes from *Ensis*, above); 3 to 4 inches (7.6-10.2cm) long. *S. rosaceus*: shell has pink cast, length/width = 5/1. Southern California and Baja California.

Spisula (Surf Clam; several species): ivory to drab white shell, often with rusty stains, yellow to brown periostracum; in general resembles *Mactra* (above) except that external and internal ligaments not separated by a plate. On exposed beaches, entire coast, intertidal to 150 feet (46m).

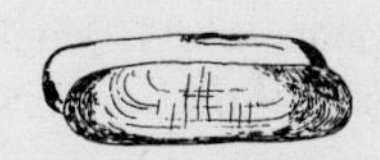

Tagelus (Jacknife Clam; several species): whitish or gray shells with rough brown periostracum, beak central, dorsal and ventral margins parallel and straight; up to 4 inches (10.7cm). Common intertidally and shallow subtidal in marshes and bays, southern California and Baja California. *T. californianus* illustrated.

Tellina (several species): shell generally white and glossy, sometimes with pink or yellow cast; elongate oval outline, anterior rounded, posterior angular. Entire coast, intertidal to 200 feet (61m) or more. *T. carpenteri* illustrated.

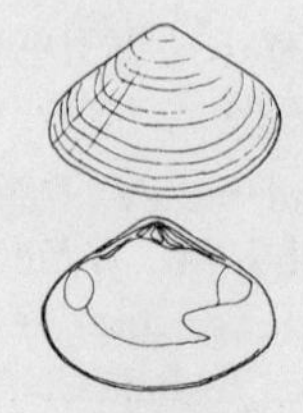

Tivela stultorum (Pismo Clam): shell dull white to light chocolate, glossy, thick, trigonal, highly symmetrical, sometimes striped radially; up to 6 inches (15cm) across. Central California to Baja California, intertidal to 10 feet (3m).

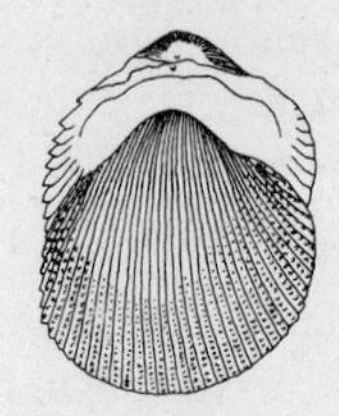

Trachycardium quadragenarium (40-ribbed Cockle): yellow to light brown shell, ribs near margins have small spines; 6 inches (15cm) across. Southern California to Baja California, intertidal to 450 feet (137m).

Tresus (*Schizothaerus*) *nuttalli* (Gaper Clam): drab white oval shell, gray periostracum, beak near anterior, moderate concentric ridging; up to 10 inches (25.4cm). Entire coast, intertidal to 120 feet (37m).

Yoldia (several species): glossy greenish to tan to brown shell, thin compressed valves; rounded anteriorly and projecting to acute angle posteriorly, ligament internal, numerous hinge teeth. *Y. cooperi*: conspicuous concentric ridges; 3 inches (7.6cm) long. Central and southern California. *Y. limatula*: faint concentric ridges; 2½ inches (6.4cm). Entire coast; genus ranges from intertidal to 6000 feet (1830m).

IV. NESTLING OR FREE-LIVING FORMS

Aequipecten (*Pecten*) *circularis* (Speckled Scallop): yellow to reddish-brown, usually mottled with light coloring; ribs smooth and even, valves strongly convex; 2 inches (5.1cm). Northern California to Baja California, intertidal to 450 feet (137m).

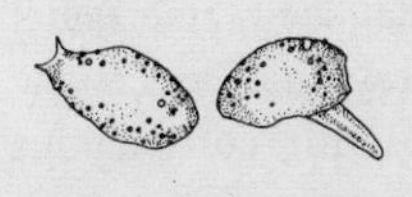

Chlamydoconcha orcutti: shell small and internal; translucent soft body, walnut-shaped; 1 inch (2.5cm) long. Beneath flat rocks and ledges, southern California, intertidal to 90 feet (27m).

Chlamys (*Pecten*) (Scallop): whitish, yellow, orange, or reddish shell, sometimes mottled; ear on anterior side longer than on posterior. *C. hastatus* (*hericia*): main ribs of left (darker) valve rough with scales or spines. *C. hindsii*: main ribs of left valve smoothish, spines tiny, right valve flatter than left. Entire coast, intertidal to 60 feet (18m) or more.

Cumingia california (*lamellosa*): white to grayish shell, trigonal, conspicuous concentric ridges, divided ligament; 1 inch (2.5cm) across. In crevices and holes, entire coast, intertidal to 150 feet (46m).

Diplodonta orbella: occasionally lying on bottom, exposed from sandy dwellings. See Group III (above) for description.

Donax (Bean Clam): occasionally lying exposed on sand surface. See group III (above) for description.

Hiatella (*Saxicava*) *arctica*: drab white thin shell, gray periostracum, red-tipped siphons, oblong-quadrate, beaks near anterior; shell often wrinkled and distorted by habitat. In crevices, holes, and holdfasts; entire coast, intertidal to 200 feet (61m).

Hinnites multirugosus (*giganteus*) (Rock Scallop): free-living when young, but adults attached. See group I (above) for description.

Leptopecten (*Pecten*) *latiauratus* (Kelp Scallop): often free-living, but usually attached to seaweed. See group I (above) for description.

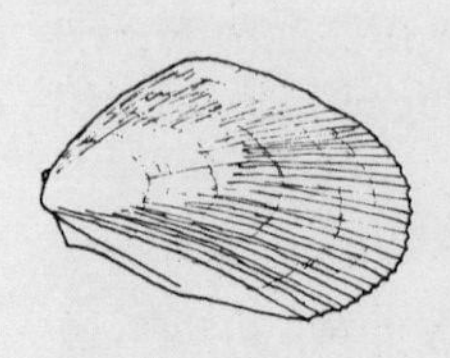

Lima hemphilli (*dehiscens*) (File Shell): whitish oblique-elliptical shell, fine radial ribs crisscrossed by very fine concentric striations; pink body with conspicuous tentacles usually protrudes from shell. Under stones, central California to Baja California, intertidal to 200 feet (61m) (Plate 4f).

Lyonsia californica: pearly shell, gray periostracum, fine radial lines, elongate; bulges anteriorly, posteriorly tapers and curves; 1 inch (2.5cm) long. Entire coast, intertidal to 40 feet (12m).

Modiolus (*Volsella*) (Horse Mussel): sometimes unattached, living at surface of sediments; more commonly attached. See group I (above) for description.

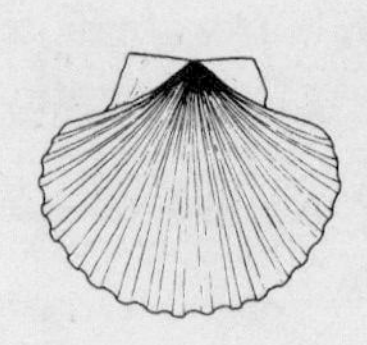

Pecten diegensis (San Diego Scallop): yellow to red shell, ribbing smooth; left valve flat, right valve only moderately concave; 3 inches (7.6cm) across. Central California to Baja California, 60 to 450 feet (18-137m).

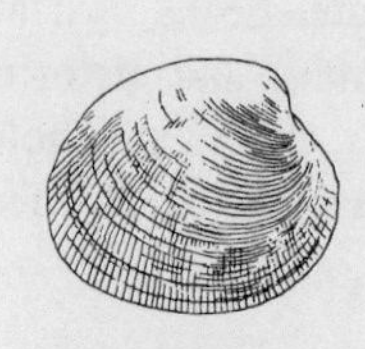

Protothaca (*Paphia, Venerupis*) (Little-neck Clam): shell gray to rusty brown, sometimes mottled, oval outline; radial and concentric ridges form network. *P. staminea* (illustrated): radial ridges more pronounced than concentric; 3 inches (7.6cm). *P. tenerrima*: concentric ridges more pronounced than radial; 4 inches (10.2cm). Entire coast, intertidal to more than 60 feet (18m).

Semele (several species): roughly circular shell, concentric ridges; ligament partly internal and extending diagonally across hinge. *S. decisa* (illustrated): thick brown shell, interior margin tinged purple; 3 inches (7.6cm) across. Southern California and Baja California, 15 to 60 feet (5-18m).

Trachycardium quadragenarium (40-ribbed Cockle): sometimes on sand surface, but also burrows. See group III (above) for description.

CLASS CEPHALOPODA–OCTOPUSES

Octopuses are the most intelligent of invertebrates. California species are not aggressive and must be pursued to collect them. They are able to bite, although they do so only rarely. Nonetheless it is wise to wear gloves when handling them, since they carry a venom. Only one genus is common in California waters at diving depths.

Octopus (*Polypus*) (Devil Fish): color varies to blend with background; soft body, 8 arms with numerous suckers. *O. bimaculatus* and *O. bimaculoides* are indistinguishable in the field; both display conspicuous round dark spot below each eye, and tentacle span up to 2 feet (61cm). Southern California and Baja California, intertidal to 80 feet (24m). *O. dofleini* (probably identical with *O. hongkongensis*): lacking spots below eyes (Plate 8d). *O. dofleini* is much redder than the other two species (not perceptible underwater, however); with tentacle span up to 16 feet (5m) but rarely seen larger than 2 feet (61cm) at diving depths off California. Entire coast, cold water, intertidal to more than 200 feet (61m).

THE PELAGIC MOLLUSKS

A few pelagic mollusks occur in California waters. Like many other pelagic animals, they are seen only sporadically under special circumstances (during plankton blooms, spawning periods, etc.); at such times they may be present in huge numbers. Some of the pelagic mollusks are always around, but avoid humans so successfully they appear scarce (e.g., squid). Others, such as pelagic gastropods, are indeed rare most of the time, but large populations may occasionally be carried close to shore by oceanic currents.

PELAGIC GASTROPODS

Atlanta (several species; *A. peroni* probably most common): fragile colorless translucent shell, coiled in a single plane; outer lip notched, keel around periphery; 1 inch (2.5cm) across. Entire coast, at or near surface.

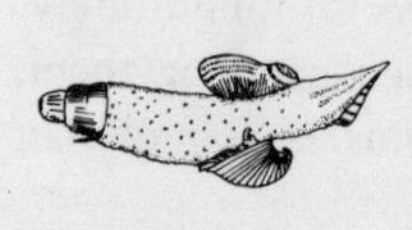

Carinaria cithara: colorless translucent shell, compressed laterally, with a double keel; located inconspicuously below jelly-like body, prominent fin uppermost; up to 10 inches (25.4cm) long. Entire coast, near surface.

Janthina (several species; *J. exigua* commonest): violet to dark purple thin fragile shell with notched outer lip; ¾-inch (1.8cm) high. Exudes gelatinous raft for buoyancy. Entire coast, near surface.

PELAGIC CEPHALOPODS

Loligo opalescens (Squid): color varies continuously from translucent white to irridescent red; 10 arms with 2 rows of sucking discs, elongate tapering body, 2 posterior fins up to 10 inches (25.4cm). Entire coast, surface to 200 feet (61m) (Plate 8f).

Chapter 11

MARINE ANIMALS: ARTHROPOD CRUSTACEA, ECHINODERMS, AND TUNICATES

THE PELAGIC CRUSTACEA (CLASS)

Pelagic crustacea are numerous and extremely important in the biological economy of the ocean. Most are quite small, requiring microscopic examination for identification and therefore beyond the scope of this book. Great schools of these creatures are encountered occasionally while diving, invariably arousing interest as to what they may be. It is usually possible to identify the major group involved while looking through a facemask; hence we will describe the two commonest forms, copepods and mysids.

Additionally, in California waters, there are pelagic barnacles and decapods that swarm in great numbers from time to time and then may not be seen for years. There are also parasitic and commensal crustaceans on many of the larger pelagic animals such as fishes, whales, jellyfishes, and tunicates. Parasitic forms include copepods and isopods, while amphipods and barnacles are frequently commensal.

SUBCLASS COPEPODA

Copepods are quite small, ranging from microscopic to about ¼-inch (0.6cm). Most are transparent and are seen darting here and there as bright suspended specks, caught by glancing sunlight. A few are dark-colored and can be detected against light backgrounds such as young kelp blades. The tremendous numbers of copepods compensate for their tiny sizes, and they are important food sources for many filter-feeding animals.

ORDER MYSIDACEA–OPOSSUM SHRIMPS

Mysids are substantially larger than copepods,commonly ranging from ½-inch to 2 inches (1.3-5.1cm) in the scuba zone. The body is streamlined, the eyes are usually prominent and dark, and color is variable, from transparent to matching whatever the typical background may be. Superficially, mysids resemble decapod shrimps, although this may be difficult to discern in a swimming school. Mysid swarms occur at all depths in the scuba zone, and may be so dense as to obscure vision. A few mysids associate closely with the large kelps and can be seen resting on or swimming among the fronts.

PELAGIC SUBCLASS CIRRIPEDIA–BARNACLES

Lepas (Gooseneck Barnacles): dark leathery stalk surmounted by enlarged capitulum, surrounded by shells or plates; length usually 1 or 2 inches (2.5 or 5.1cm). *L. anatifera* (illustrated): calcareous bluish-white shells with faint radial and concentric lines; attached to logs and debris. Entire coast. *L. ansifera*: calcareous plates, radially grooved. Entire coast. *L. fascicularis*: plates not calcareous but paper-like; attached on drift seaweed or free-floating. Northern and central California. *L. hilli*: calcareous smooth plates, top of stalk pale or orange. Northern and central California.

PELAGIC ORDER DECAPODA–SHRIMPS AND CRABS

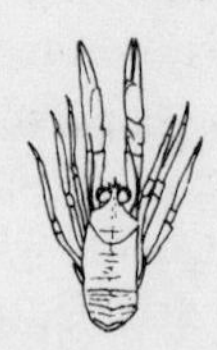

Pleuroncodes planipes (Squat Lobster): bright red, pincer arms long; resembles lobster but with greatly reduced abdomen and tail; 3 inches (7.6cm). Southern California and Baja California.

Portunus xantusii (Swimming Crab): also found on the bottom. Described under Brachyuran Decapods in Benthic Crustacea section (below).

BENTHIC CRUSTACEANS (CLASS)

California waters contain a great many species of benthic crustaceans. (Crustacea are the principal marine class in the Phylum Arthropoda.) Most are quite small, and although they may be very numerous we shall have to ignore them, since identification usually requires microscopic techniques. Also, they typically inhabit labyrinth-like environments such as sponges, holdfasts, or shell masses, and recovery of such specimens requires painstaking laboratory dissection of the environmental home, carefully picking out the tiny inhabitants. We will confine our discussion to larger forms, perceived and collected fairly easily by a diver.

Crustaceans possess hard external shells and jointed arms, legs, or appendages. One group, the Cirripedia or barnacles, is attached; but the other species we shall consider are fairly mobile, although a few live sedentarily in burrows. They occur in sand and on rocks, in bays and along open coasts. A few of the larger crabs could cause injury with their pincers but rarely do so, because their slow movements are easily avoided. The sharp edges of barnacle shells cut flesh readily and should be approached cautiously.

SUBCLASS CIRRIPEDIA–THE BARNACLES

The protective shell of barnacles is formed of several plates, either fused or separated. Number and appearance of the plates is used for identification. Be careful, therefore, to avoid injury to the plates when chiseling a barnacle loose.

Only three genera of barnacles occur commonly in subtidal California waters, and most species belong to a single genus, *Balanus*. (Another barnacle genus, *Lepas*, is included under Pelagic Crustacea, above.) We have therefore varied here from our usual alphabetic presentation. The genera are first distinguished, following which species of *Balanus* are discussed.

Tetraclita squamosa: red shell, deeply ridged, of 4 fused plates.

Scalpellum osseum: plates clearly separated, situated at end of a stalk.

Balanus (several species): 6 plates, margins may be obscured. *B. tintinnabulum* (illustrated): pink with white vertical striping, walls smooth or finely ridged; up to 2 inches (5.1cm) in diameter. Central California to Baja California, intertidal to 40 feet (12m). *B. cariosus*: drab white, finely ridged, with basal portions of ridges extended as pointed filaments; up to 2 inches (5.1cm) in diameter, usually much smaller. When crowded, shell elongates, resembling a hexagonal pencil. Entire coast, intertidal to at least 50 feet (15m). *B. crenatus*: white, smooth or grooved walls, base thin and calcareous; width up to ½-inch (1.3cm). Often on undersides of rocks; entire coast, intertidal to 50 feet (15m). *B. nubilis*: drab white, often encrusted, a few irregular ridges, mantle with vivid blue and red bands; usually large, 3 inches (7.6cm) in diameter. Entire coast, intertidal to 100 feet (30.5m) (Plate 7a). There are many other *Balanus* species also, requiring laboratory dissection for identification.

ORDER STOMATOPODA–MANTIS SHRIMPS

Only one common stomatopod is known in California subtidal waters, and it occurs rather deeply.

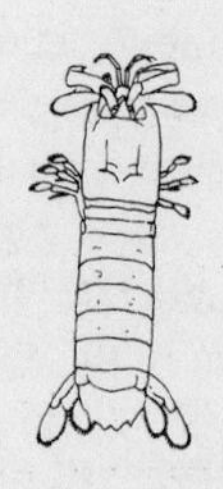

Pseudosquilla bigelowi: gray body with bright yellow and blue markings on the frontal appendages; 8 inches (20.3cm) long. It digs burrows the size of gopher holes, 2 or 3 feet (61-91cm) deep. Animals may roam in the open in early morning or late afternoon when daylight is waning. *Caution*: one frontal appendage is scalpel-like, can inflict deep gashes in flesh. Entire coast, 30 to at least 100 feet (9-30.5m).

ORDER ISOPODA–PILLBUGS, GRIBBLES, etc.

Isopods are flattened dorso-ventrally (as if they had been stepped on). Many species infest seaweed holdfasts, and all are small creatures. One genus, *Limnoria*, burrows in woody structures and causes great damage to piers and docks. A few species are sufficiently large or important in their habitat to deserve mention.

Cirolana harfordi: light tan or gray with black spotting, almost translucent, resembles the common terrestrial "sow-bug"; ¼- to ½-inch (0.6-1.3cm), often common on undersides of rocks. Entire coast, intertidal to 60 feet.

Idothea (several species): markedly flattened and elongate, resembles a shortened centipede. *I.* (*Pentidotea*) *resecata* (illustrated) (Kelp Isopod): greenish-yellow, posterior margin concave, forming two points; 2 inches (5.1cm) long. Entire coast, intertidal to 60 feet (18m). *I. rufescens*: similar but reddish, and front more pointed. Entire coast, intertidal to 60 feet (18m).

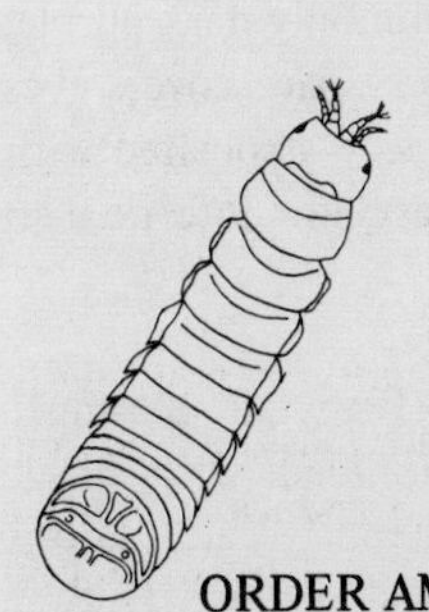

Limnoria (gribble; several species): burrow in wood and seaweed stripes; generally small, about 1/8-inch (0.3cm). *L. algarum* (illustrated) ranges to 110 feet (34m), in basal parts of brown seaweeds.

ORDER AMPHIPODA–SEAWEED FLEAS, SKELETON SHRIMPS, etc.

Most amphipods are flattened laterally (as if they had been pinched). They are small like isopods. They infest complex structures such as holdfasts and also occur in the upper portions of the more bushy algae. They are often not detected until a shell or clump of seaweed is placed in an aquarium, and then the tiny flealike amphipods are seen crawling or swimming in the vicinity.

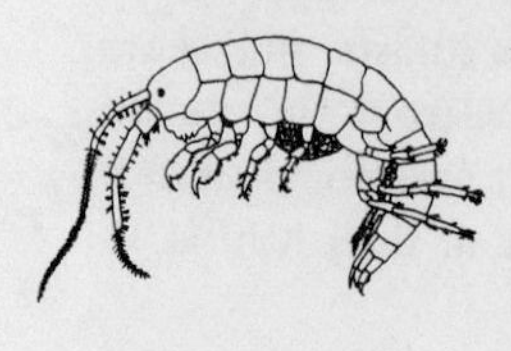

Ampithoë humeralis (Kelp Curler): yellow-green, antennae long; ½-inch (1.3cm). Inhabits crevices formed by bending and gluing edge of kelp blade back against itself. Entire coast, surface to 50 feet (15m).

Caprella (Skeleton Shrimps; several species): elongate sticklike body, clings tenaciously by recurved hooks on legs, up to 1 inch (2.5cm) long. Microscopic examination necessary to distinguish species. Entire coast, at all scuba depths. *C. aequilibria* illustrated.

ORDER DECAPODA, SECTION MACRURA– LOBSTER, TRUE SHRIMPS

Betaeus harfordi (Abalone Shrimp): commonly dark purple, also red and green; conspicuous pincers, carapace covers eyes anteriorly; 1 inch (2.5cm) long. Often closely associated with abalone and large sea urchins. Entire coast, low intertidal to 130 feet (40m).

Crago nigricauda (Blacktail Shrimp): light background sprinkled with black dots, conspicuous antennal scales, dorsal ridge on 5th abdominal segment; 2 to 3 inches (5.1-7.6cm). Burrows in sand; entire coast, low intertidal to 60 feet (18m).

Crangon (*Alpheus*) *dentipes* (Pistol Shrimp or Snapping Shrimp): green, sometimes with brown or orange mottling; one pincer usually very large and used for making sharp clicking noise; 2 inches (5.1cm). Central California to Baja California, intertidal to 100 feet (30.5m).

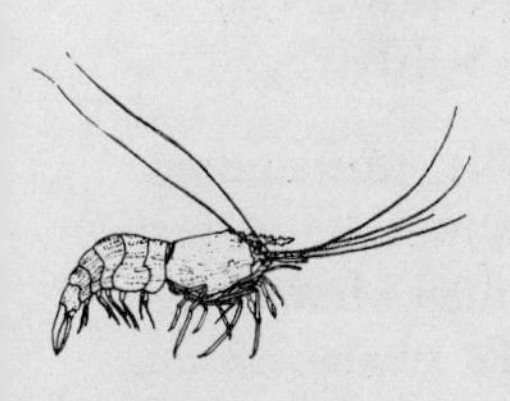

Hippolysmata californica: semi-transparent with red striping, long red antennae and antennules; 2 inches (5.1cm). Aggregates with morays in crevices. Southern California and Baja California, low intertidal to 130 feet (40m) (Plate 2e).

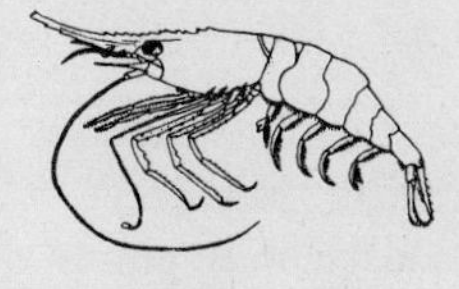

Pandalus (Coon-striped Shrimp): conspicuous rostrum longer than carapace, spines of rostrum continue posteriorly forming ridge on carapace, 2nd pair of legs unequal. Low intertidal to 100 feet (30.5m) or more. *P. danae* (illustrated): rostrum slightly curved upward; 4 inches (10.2cm). Northern and central California. *P. gurneyi*: rostrum strongly curved upward; 3 inches (7.6cm). Southern California.

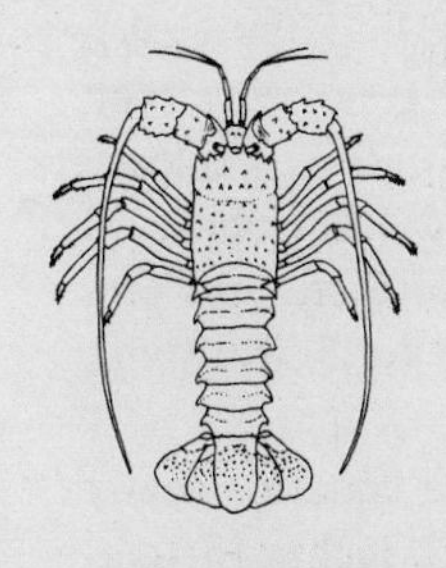

Panulirus interruptus (Spiny Lobster): dull red, legs with green marking, carapace and base of antennae studded with spines; up to 36 inches (91cm) long. Southern California and Baja California, low intertidal to more than 200 feet (61m).

Spirontocaris (Broken-back Shrimp): often semi-transparent, with bright green, red, or orange markings; abdomen after third segment bends downward at right angle to rest of body, segment behind pincer on second pair of legs divided into 7 annulations; 1 inch (2.5cm). Several species, requiring microscope to distinguish. Intertidal to at least 130 feet (40m). *S. palpator* illustrated.

ORDER DECAPODA, SECTION ANAMURA–HERMIT CRABS, SAND CRABS, GHOST SHRIMPS, etc.

The Anomura may superficially resemble true shrimps (Macrura) or true crabs (Brachyura) or neither, and at first sight appear to be a motley group. Fundamentally, however, it is clear that they are closely related. Crablike anomurans can be distinguished from true crabs by the fact that the eyes lie between the antennae; in true crabs the antennae lie between the eyes.

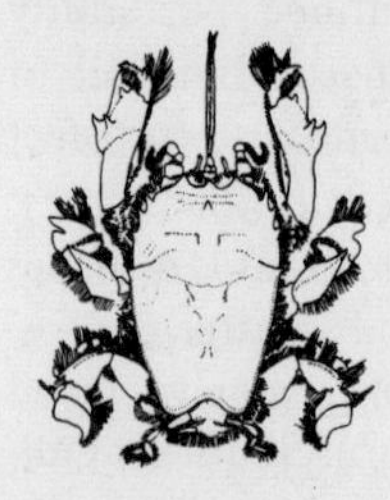

Blepharipoda occidentalis (Spiny Sand Crab): cream-colored with blue-gray tinge, flattened body, conspicuous pincers, spines along anterior carapace margin; 3 inches (7.6cm). Burrows in sand; central California to Baja California, low intertidal to 100 feet (30.5m).

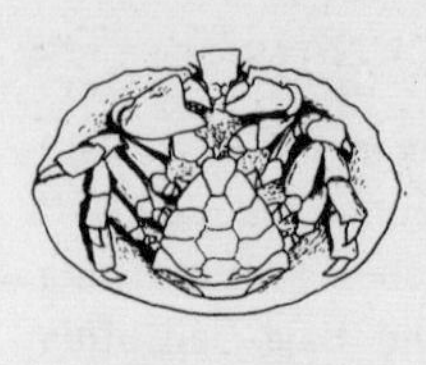

Cryptolithodes sitchensis (Sitka Crab): bright red (males) or gray with pink marking (females); oval outline, carapace hides legs when viewed from above; 2 inches (5.1cm) wide. Northern and central California, intertidal to 20 feet (6m).

Holopagurus pilosus (Hermit Crab): chelae equal or nearly equal, on sediment surfaces or burrowing; 2 inches (5.1cm) long. Central California to Baja California, intertidal to 200 feet (61m).

Pachycheles (several species): body moderately flattened, chelae massive and stubby, inner segments as broad as they are long; ½-inch (1.3cm) wide. Entire coast. *P. pubescens* (illustrated): telson with 7 plates, chelae hairy. Intertidal to 60 feet (18m). *P. rudis*: telson with 5

plates, chelae tuberculated. Intertidal to 40 feet (12m).

Paguristes (Hermit Crab): anterior abdominal segments with sexual appendages (distinguishes from *Pagurus*, below. *P. bakeri*: chelae spiny, inner margin straight. Southern California and Baja California, intertidal to 80 feet (24m). *P. turgidus*: chelae rounded and inflated. Entire coast, intertidal to 100 feet (30.5m). *P. ulreyi*: chelae rounded and oblong. Central and southern California, intertidal to 80 feet (24m).

Pagurus (Hermit Crab): lacking sexual appendages on anterior abdominal segments (distinguishes from *Paguristes*, above). *P. granosimanus*: anterior carapace margin round. Entire coast. *P. hemphilli*: anterior carapace margin pointed. Northern and central California, intertidal to 50 feet (15m). *P. hirsutiusculus* (illustrated): anterior carapace wider than long. Entire coast, intertidal to 100 feet (30.5m).

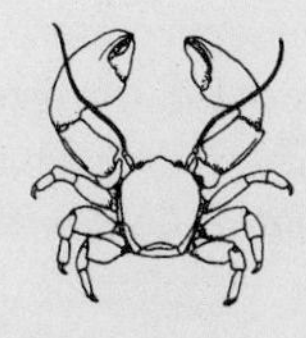

Petrolisthes (Porcelain Crab): brown to green, body strongly flattened, circular outline. Lab study necessary for identification to species. Entire coast, intertidal to 150 feet (46m). *P. cinctipes* illustrated.

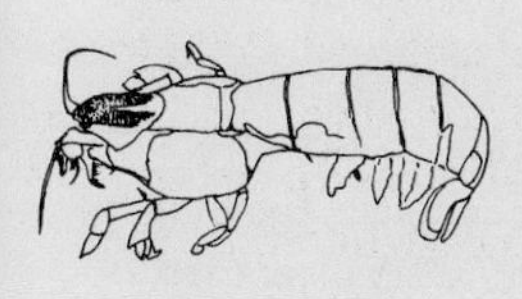

Upogebia pugettensis (Mud or Ghost Shrimp): bluish-gray; movable "thumb" of pincer is considerably shorter than matching "finger." Burrows in fine sediments; entire coast, intertidal to about 20 feet (6m).

ORDER DECAPODA, SECTION BRACHYURA–TRUE CRABS

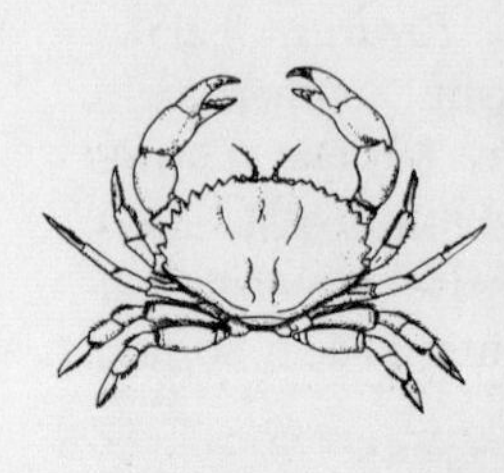

Cancer (Rock Crab): pink, purple, brown or red; body flattened, oval, anterior carapace margin with conspicuous teeth, 5 frontal teeth between eyes. *C. antennarius* (illustrated): widest at 8th tooth, red spots on ventral side. *C. anthonyi*: carapace widest at 9th anterolateral tooth, ventral side whitish, not hairy dorsally. *C. jordani*: widest at 9th tooth, hairy dorsally. *C. magister*: carapace widest at 10th anterolateral tooth, pincers white-tipped. *C. productus*: 5 frontal teeth, about equal size (unequal in other species). Entire coast (except *C. antennarius*, southern California and Baja California), intertidal to 50 feet (15m) or deeper.

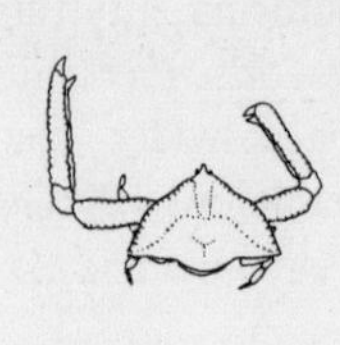

Heterocrypta occidentalis (Elbow Crab): pink to blue-gray, carapace triangular, conceals legs, pincer arms much longer than legs; 1 inch (2.5cm). On sediment bottoms; central California to Baja California, low intertidal to 40 feet (12m).

Lophopanopeus (several species): anterior carapace margin notched in middle, chelae black, carapace trapezoidal, widest anteriorly; ½-inch to 1 inch (1.3-2.5cm) wide. *L. diegensis*: pincer arm tuberculated. Central and southern California, intertidal to 50 feet (15m). *L. heathi*: pincer arm smooth. Northern and central California. *L. leucomanus*: pincer arm pitted and ridged. Central California to Baja California, intertidal to 50 feet (15m).

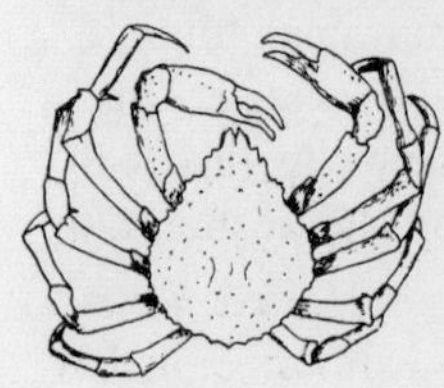

Loxorhyncus (several species): round to pear-shaped body tapering to two prominent points anteriorly, carapace tuberculated. *L. crispatus* (Spider Crab): carapace with 9 to 12 large blunt tubercles; 3

inches (7.6cm) wide. *L. grandis* (Sheep Crab) (illustrated): carapace with many pointed tubercles; 6 inches (15cm) wide. Central California to Baja California, low intertidal (occasionally) to 400 feet (120m).

Mimulus foliatus: tan, purple, reddish; outline roughly octagonal (counting bifurcated rostrum as one side); 1 inch (2.5cm) wide. Entire coast, intertidal to 400 feet (120m).

Paraxanthias taylori (Lumpy Crab): dull red, conspicuous rounded tubercles on pincer arms and anterior carapace, legs with short bristles; 1½ inches (3.8cm). Central California to Baja California, intertidal to 150 feet (46m).

Pelia tumida (Dwarf Crab): drab to orange, adults without spines, hairy; carapace pear-shaped, elongating to two small points anteriorly; ¾-inch (1.8cm) long. Usually covered with sponges and algae. Southern California and Baja California, intertidal to 300 feet (91m).

Pilumnus spinohirsutus (Hairy Crab): sandy to brown, thick cover of bristles; 1 inch (2.5cm). Southern California and Baja California, intertidal to 40 feet (12m).

Pinnixa (Pea Crab): oval to elliptical carapace, flattened legs; 3rd pair of legs is largest. Several species, all smaller than 1 inch (2.5cm). Typically commensal with mollusks or tube-building polychaetes. *P. longipes* (illustrated): carapace elongate laterally, 3rd legs each larger than body. Entire coast, intertidal to 40 feet (12m).

Portunus xantusii (Swimming Crab): blue-gray, carapace roughly elliptical with numerous marginal teeth anteriorly, terminating in two long lateral spines, fourth leg pair modified as paddles; 3 inches

(7.6cm). In bays and on sediment bottoms, southern California and Baja California, intertidal to at least 60 feet (18m).

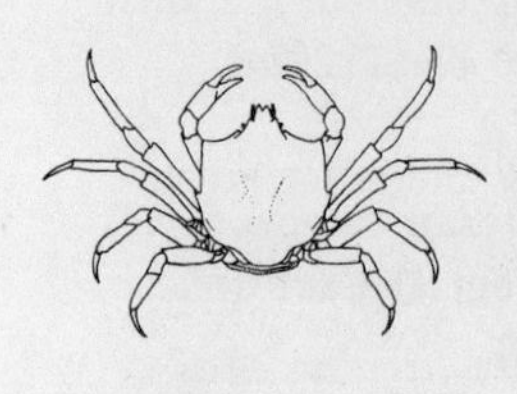

Pugettia (*Epialtus*) (Kelp Crab): outline polygonal with prominent teeth at corners, rostrum with 2 teeth. *P. dalli*: outline triangular, tooth just behind eye not in same plane as other carapace teeth, carapace lumpy; ½-inch (1.3cm). Southern California and Baja California, intertidal to 300 feet (91m). *P. producta* (illustrated): olive to brown, carapace smooth; 4 inches (10.2cm). Entire coast, intertidal to 250 feet (76m). *P. richi*: reddish, carapace lumpy, outline roughly diamond-shaped; 1½ inches (3.8cm). Entire coast, intertidal to 300 feet (91m).

Pyromaia (*Inachoides*) *tuberculata* (Decorator Crab): drab, usually covered with encrustations, carapace triangular to circular, lumpy, rostrum a single spine, flanked by spines behind eyes; ½-inch (1.3cm). Entire coast, intertidal to 270 feet (82m).

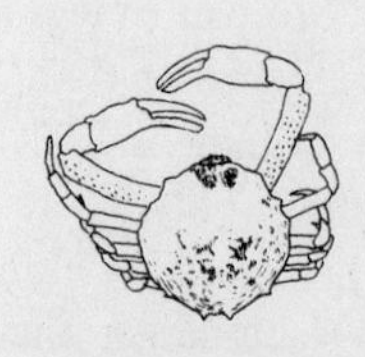

Randallia ornata (Purple Globe Crab): whitish, mottled with red-purple; carapace hemispherical, rostrum almost absent; 2 inches (5.1cm). Entire coast, intertidal to at least 40 feet (12m).

Scyra acutifrons (Sharpnose Crab): drab gray or tan, outline triangular, carapace with very irregular tubercles, rostrum with two pear-shaped spines; 2 inches (5.1cm) long. Entire coast, intertidal to 300 feet (91m).

Taliepus (*Epialtus*) *nuttalli* (Red Kelp Crab): bright to dark red or purple, carapace smooth, outline circular, rostrum slightly bifurcated; 4 inches (10.2cm). Southern California and Baja California, intertidal to 300 feet (91m).

PHYLUM ECHINODERMATA

The common major classes of echinoderms comprise animal types that differ greatly in superficial appearance: sea stars (starfish), urchins, brittle stars (also called serpent stars), and sea cucumbers. Novices are often surprised to find that a wriggling brittle starfish is related to a sluggish wormlike sea cucumber. Except for a few brittle stars, echinoderms are slow-moving or sedentary. Nonetheless, they may be very important in their communities. Sea stars are often significant predators. Urchin swarms frequently control much territory by their intensive grazing activities. Brittle stars and sea cucumbers clean up detritus and serve an important janitorial function for the community.

CLASS ASTEROIDEA–THE SEA STARS

Astrometis sertulifera: usually green-brown, with blue-banded, red-tipped spines; sometimes yellow, orange, purple, or mottled–particularly in deep or cold water; up to 10 inches (25.4cm) across. Southern California and Baja California, intertidal to 60 feet (18m).

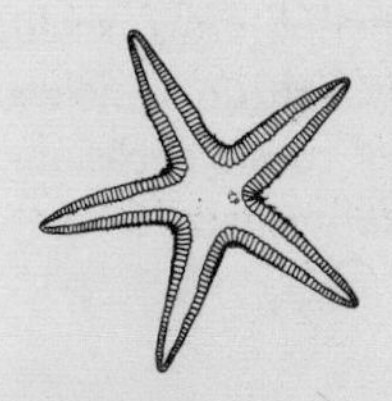

Astropecten (Sand Star): gray or tan, very symmetrical, margins grooved to form conspicuous plates; sometimes bluish, prominent lateral spines. *A. armatus* (illustrated): 8 inches (20.3cm). Intertidal to 40 feet (12m). *A. verrilli*: 4 inches

(10.2cm). From 2 to 100 feet (0.6-30.5m). Both species on sediment bottoms in southern California and Baja California.

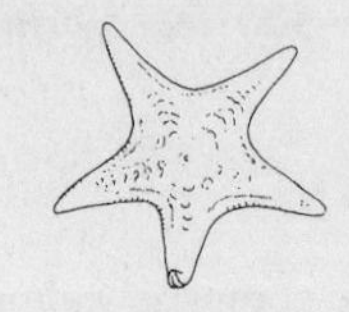

Dermasterias imbricata (Leather Star): mottled, drab yellow or orange background with indented red patches; smooth, base of arms broad; up to 8 inches (20.3cm). Entire coast, intertidal to 100 feet (30.5m).

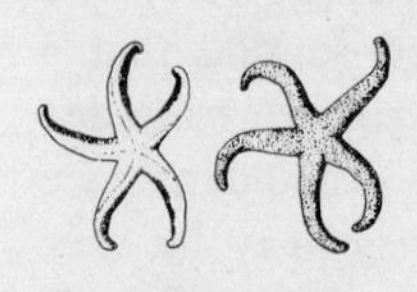

Henricia leviuscula: yellow, orange, or red; long narrow arms, fairly smooth; up to 10 inches (25.4cm). Entire coast, cold water, intertidal to 100 feet (30.5m) (Plate 2f).

Leptasterias (several species): narrow arms, moderately rough, small animals. *L. aequalis:* white, pink, red, sometimes mottled; 2½ inches (6.4cm). *L. pusilla* (illustrated): green to gray, 1 inch (2.5cm) across. Shallow. Entire coast in cold water low intertidal to 100 feet (30.5m).

Linckia columbiae: red mottled with gray, smooth, arms narrow, almost circular in cross-section; 4 inches (10.2cm). Southern California and Baja California, intertidal to 60 feet (18m).

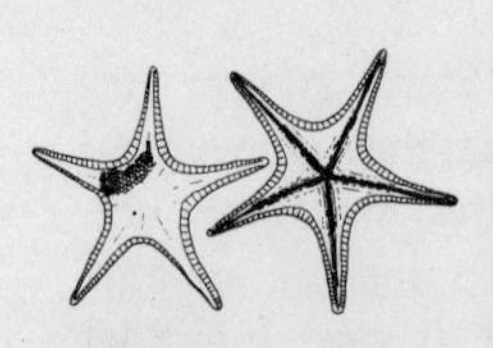

Mediaster aequalis: bright red, broad center with sharply tapering arms, small marginal plates, slightly rough; 4 inches (10.2cm) across. Entire coast, occasionally shallow in cold water, usually 60 feet (18m) or deeper.

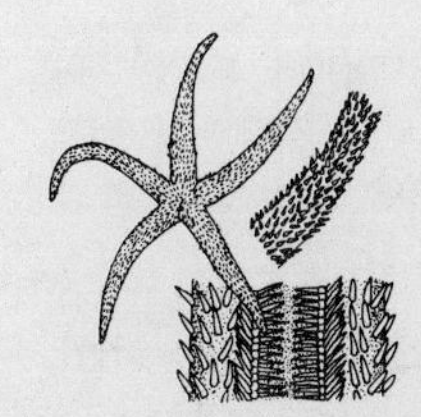

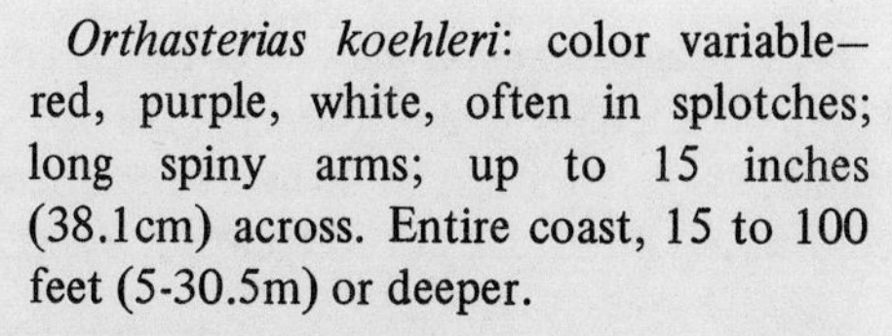

Orthasterias koehleri: color variable–red, purple, white, often in splotches; long spiny arms; up to 15 inches (38.1cm) across. Entire coast, 15 to 100 feet (5-30.5m) or deeper.

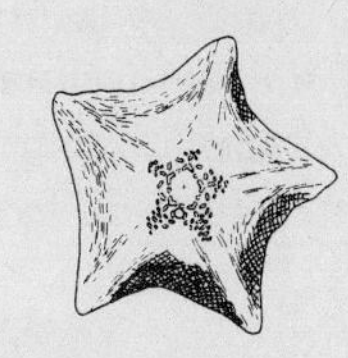

Patiria miniata (Bat Star): yellow, orange, red, or purple, as uniform color or mottled; smoothish, base of arms broad, giving webbed appearance; up to 6 inches (15.2cm). On rock and sediment bottoms, entire coast, intertidal to 100 feet (30.5m) (Plate 7b).

Petalaster (*Luidia*) *foliata*: gray, slightly mottled or spotted, smooth, narrow arms, resembles *Astropecten* (above) but lacks marginal plates; up to 16 inches (40.6cm). On sediment bottoms in cold water, entire coast, 40 to more than 100 feet (12-30.5m).

Pisaster brevispinus (Pink Star): pale pink, short spines gathered into clusters; up to 24 inches (61cm). Usually on sediment bottom; entire coast, low intertidal to over 300 feet (91.5m).

Pisaster giganteus (Giant Star): random white stubby spines surrounded by blue plaques, interspersed by a yellow and dark blue background; 22 inches (56cm). Entire coast, intertidal to 80 feet (24.5m) (Plate 7a).

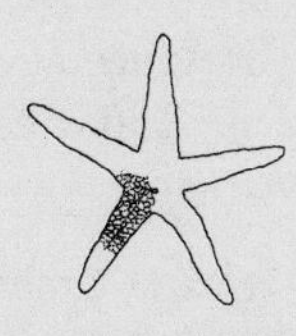

Pisaster ochraceus (Ochre Star): yellow, orange, purple, brown, red, stubby spines arranged in a network; 16 inches (40.6cm). Entire coast, intertidal to 30 feet (9m).

Pteraster tesselatus arcuatus: disc 1 to 2 inches (2.5-5.1cm) thick, very broad rays (arms) turning up at tips, mouth contains 5 pointed teeth; 6 inches (15.2cm) across. Central California, 50 feet (15m) or deeper.

Pycnopodia helianthoides (Sun Star): commonly purple or blue, occasionally yellow, orange, or red; soft, with stubby spines, 15 to 24 arms; 24 inches (61cm) across. Entire coast in cold water, intertidal to 100 feet (30.5m) (Plate 7b).

Solaster dawsoni (Sun Star): red, orange, yellow, gray, or blue, uniform or striped; smooth, 8 to 13 arms; 14 inches (35.6cm) across. Northern and central California, occasionally southward; sometimes intertidal, more often 10 to 50 feet (3-15m).

CLASS ECHINOIDEA–THE URCHINS

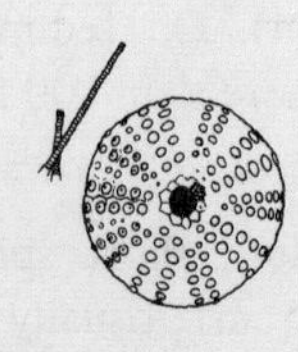

Centrostephanus coronatus: black or dark purple, spines twice the body diameter and serrated; test up to 2½ inches (6.4cm). (To distinguish from *Strongylocentrotus* [below] underwater, feel serrations *gently*.) Abundant around islands, southern California and Baja California, 20 to 100 feet (6-30.5m).

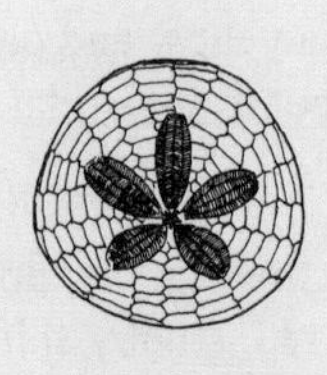

Dendraster (Sand Dollar): circular test, thin like a coin, 3 inches (7.6cm) in diameter; fine spines when alive. *D. excentricus* (illustrated): dark purple. Buries partially. Entire coast, intertidal to 50 feet (15m), sedimentary bottoms. *D. laevis*: whitish or light green. Buries completely, 30 to 100 feet (9-30.5m).

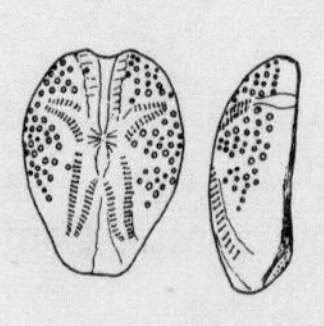

Lovenia cordiformis (Heart Urchin): grayish-white, fragile; test walnut-shaped, upper spines recurved; 1½ inches (3.8cm) long. Buries in sediment; southern California and Baja California, low intertidal to 60 feet (18m) or more.

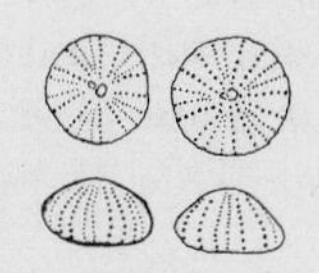

Lytechinus anamesus (White Urchin) (illustrated): white, variable reddish mottling or spots, 1 inch (2.5cm) in diameter. Southern California and Baja California, shallow subtidal to more than 100 feet (30.5m). Related yellowish-white *L. pictus* occurs in bays.

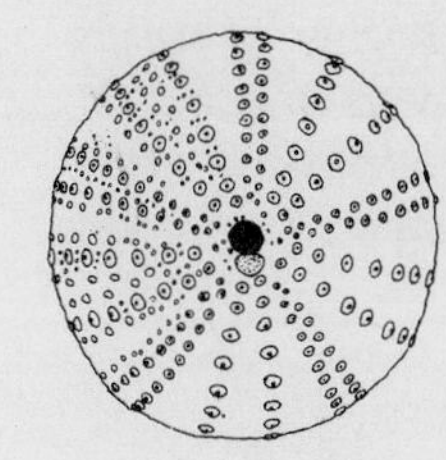

Strongylocentrotus franciscanus (Giant Urchin): dark purple or ochre, spines smooth (distinguishes from *Centrostephanus*, above); up to 6 inches (15.2cm) in diameter. Entire coast, intertidal to more than 100 feet (30.5m) (Plates 2f and 7c).

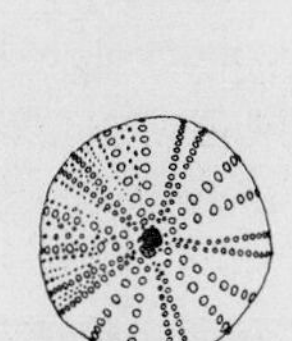

Strongylocentrotus purpuratus (Purple Urchin): adults various hues of blue or bright purple, juveniles greenish; 1 to 2 inches (2.5-5.1cm) in diameter. Common, entire coast, intertidal to 80 feet (24.5m) (Plate 8b).

CLASS HOLOTHUROIDEA–THE SEA CUCUMBERS

Cucumaria (several species): blackish or orange-red, mottled, usually seen as fuzzy circlet of many-branched tentacles protruding from crevice; 1 or 2 inches (2.5 or 5.1cm) in diameter. *C. lubrica*: black body. *C. miniata*: salmon-pink body. Entire coast in cold, clear water, intertidal to 80 feet (24.5m).

Eupentacta quinquesemita: white body, yellow tentacles; unretractile tube feet give bristly appearance. Northern and central California, intertidal and shallow subtidal.

Leptosynapta albicans: white, translucent when expanded; small whitish tentacles, longitudinal striping caused by musculature; usually 2 or 3 inches (5.1-7.6cm) long. Buries in sediment; entire coast, intertidal to 50 feet (15m).

Molpadia arenicola (Sweet Potato): yellowish, purple, or brown, often mottled; smooth, slippery, tapering sharply, lacks tube feet; up to 10 inches (25.4cm) long. Buries in sand; central California to Baja California, intertidal and shallow subtidal.

Parastichopus (several species): reddish-brown, tentacles present but often withdrawn, shape and size approximate a hot-dog bun; up to 18 inches (45.7cm) (Plate 1e). *P. californicus* has prominent pointed papillae; *P. parvimensis* numerous small tube feet. Rocky and sedimentary bottoms; entire coast, intertidal to 100 feet (30.5m) or more.

Psolus chitinoides: white or orange, flattened in cross-section, upper surface displays firm plates, tube feet prominent beneath; tentacles when expanded are prominent, red to purple; up to 5 inches (12.7cm). Under rocks in cold water; northern and central California and Baja California, low intertidal to at least 60 feet (18m).

Thyonepsolus nutriens: red body, yellowish to purple tentacles, flattened in cross-section, tube feet on under side; up to 1 inch (2.5cm). Frequents seaweed holdfasts; entire coast, low intertidal to shallow subtidal.

CLASS OPHIUROIDEA–THE BRITTLE STARS

Amphiodia (several species): very long arms, often 10 or more times disk diameter; disk covered with fine scales, some with pointed free ends; 6 *papillae* (teeth) per jaw (5 jaws), all approximately equal size. Arms fragment easily when handled. *A. occidentalis* (illustrated, top): gray to green disk, yellow-white arms. Intertidal to 90 feet (27.5m). *A. urtica* (illustrated, bottom): disk gray or red, arms whitish. From 60 to 600 feet (18-183m). Often buried in sediment with tips of arms protruding; entire coast.

Amphipholis pugetans: black with white markings, disk with scales; 6 oral papillae per jaw, outermost pair largest; up to ¾-inch (1.9cm) armspread. Entire coast, intertidal to 250 feet (76m).

Ophioderma panamense: black to dark brown, arms banded by light rings, smooth disk, many oral papillae; up to 12-inch (30.5cm) armspread. Southern California and Baja California, intertidal to 90 feet (27.5m).

Ophionereis annulata: grayish, arms banded with dark rings, disk scaled and spotted, 3 spines per lateral arm plate, 9 oral papillae; up to 8 inches (20.3cm). Southern California and Baja California, low intertidal to 30 feet (9m).

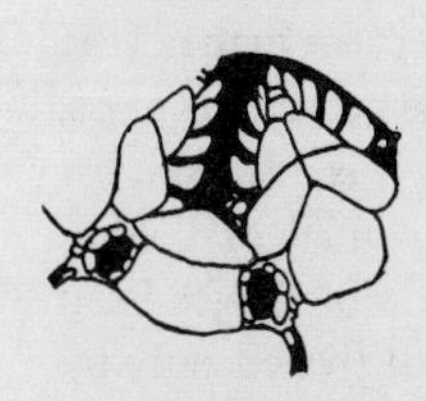

Ophioplocus esmarki: uniform pink to gray, disk scaled, numerous oral papillae, arms relatively short and flattened, upper side of arm covered by many small plates; armspread up to 3 to 4 inches (7.6-10.2cm). Central California to Baja California, intertidal to 40 feet (12m).

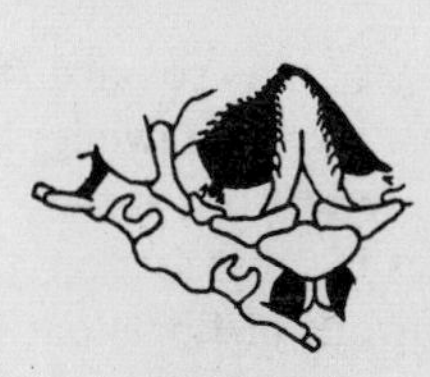

Ophiopteris papillosa: brown, dark and light banding on arms, many oral papillae, blunt stumps cover disk; arm spines conspicuous, 5 per row; up to 7 inches (17.8cm). Central California to Baja California, low intertidal to 250 feet (76m).

Ophiothrix spiculata: variable color—blue, green, orange, red, yellow, white, etc., often brilliant—disk often different from arms; spines cover disk, lacks oral papillae; arm spines conspicuous, serrated, 7 per row; up to 5 inches (12.7cm).

Common, often in aggregates, central California to Baja California, low intertidal to 250 feet (76m) (Plate 4a).

Ophiura lutkeni: gray, disk sometimes spotted, scaled, many oral papillae, upper side of disk notched and bears papillae at arm junction; armspread 3 to 4 inches (7.6-10.2cm). Sometimes swarms on sediment surfaces; entire coast, 60 to 2000 feet (18-610m).

THE BENTHIC TUNICATES (SUBPHYLUM)

Benthic tunicates are attached in one spot for all of their adult lives. They display perhaps the widest range in appearance of any major group. A few species occur as solitary individuals but most are colonial, joined at the base rather loosely by interconnections such as stolons (the social tunicates), or joined intimately by being embedded together in a matrix (the compound tunicates). The matrix is usually a gelatinous mass, but it may be so perfused with sediment that it is mistaken for a cake of mud or sand. Surface of the individual or colony may be smooth, rough, or deeply crinkled. Color ranges from drab tans and grays to brilliant orange or red, rivaled only by the brightest sponges.

It is sometimes easy to confuse sponges and colonial tunicates, since both frequently appear as inert sprawling masses. The safest way to distinguish them is to tear or cut the mass so as to expose a cross-section. Sponges in cross section vary from uniform to a rather simple meshwork of fibers or spicules, interspersed with soft tissues. Colonial tunicates are much more complex. To the naked eye the cross-section exposes many tiny elongate and often highly-colored, individual animals embedded in the matrix.

Tunicates vary as much in size as in their other characteristics. Solitary adults range from button-sized blobs up to stalked individuals 10 or more inches (25.4cm) tall. Colonial species may be small individually, but the colony sometimes spreads over several square feet (m) of rock surface. All tunicates have an incurrent and excurrent aperture, or *siphon*, for circulating seawater and this is often useful, though not infallible, in characterizing the group. Sometimes apertures are too small to be seen without a microscope, and certain colonial groups discharge seawater through a common aperture. Certain other groups also have 2 apertures. Bivalve mollusks, for example, have two apertures, but they can be distinguished by their hard shells. The covering of tunicates (i.e., the tunic) ranges from soft and delicate to tough and leathery, and a hard shell is never constructed. For field characterization, the terms solitary, social, and compound are useful; such a classification, however, places related animals in differing categories.

I. Solitary–Animals isolated or easily separable if in clumps
II. Social–Individuals joined basally
III. Compound–Animals embedded in a common matrix

I. SOLITARY

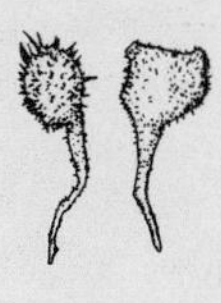

Boltenia villosa: round to oval body surmounting 1/16-inch (0.15cm) diameter stalk; free end with many hairs or spines, sometimes branching. Entire coast, low intertidal to 300 feet (91.5m).

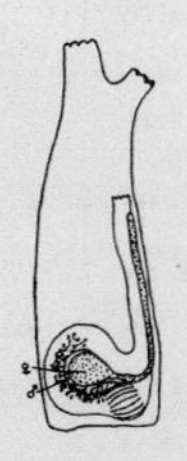

Ciona intestinalis: translucent green, elongate, soft, moderately wrinkled; attached at one end, conspicuous apertures at other end; up to 3 or 4 inches (7.6-10.2cm) long. Common in bays and harbors, entire coast, usually low intertidal and shallow subtidal.

Chelyosoma productum: translucent (young) to opaque, drab tan or yellow; heavily wrinkled, flattened top with 6 triangular plates surrounding each aperture; 1 inch (2.5cm) tall. Northern to southern California, low intertidal to 150 feet (46m).

Cnemidocarpa finmarkiensis: bright red or pink to pearly white, smooth oval body, broad basal attachment, conspicuous apertures; up to 1 inch (2.5cm) high. Inhabits rock crevices, northern and central California, low intertidal to about 100 feet (30.5m).

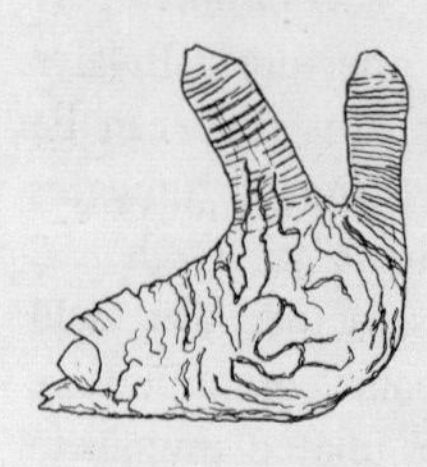

Pyura haustor: drab, opaque tunic, often covered with debris; deeply wrinkled, body cylindrical, apertures prominent, tipped with red when expanded; up to 3 inches (7.6cm) tall. Entire coast, harbors and open sea, low intertidal to 80 feet (24.5m).

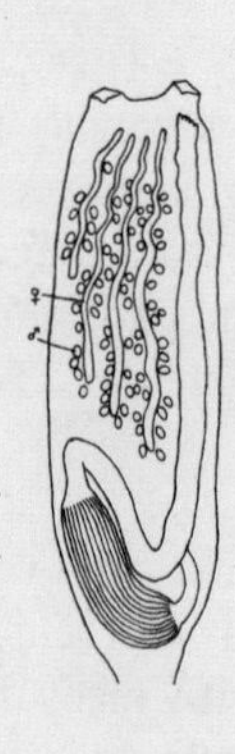

Styela (several species): tunic drab, opaque, tough, may be ridged; elongate oval body tapering to a stalk, apertures conspicuous. *S. barnharti* (illustrated): purple apertures. Common in southern California harbors. *S. montereyensis*: orange apertures, body often has pinkish cast; up to 10 inches (25.4cm) tall. Common on open rocky coasts. *S. plicata*: lacks stalk, has radiating alternate brown and yellow bands surrounding apex. In bays. Entire coast, low intertidal to more than 100 feet (30.5m).

II. SOCIAL

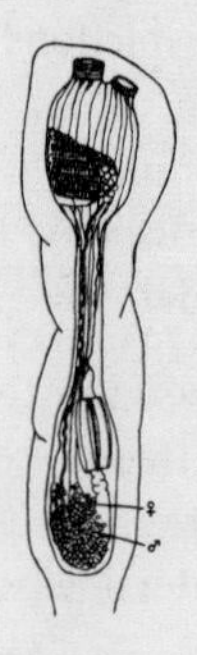

Clavelina huntsmani: transparent body revealing bright pink thorax within; club-shaped individuals, up to 1½ inches (3.8cm) tall. In crevices and overhangs, northern and central California and Channel Islands, low intertidal to 60 feet (18m) (Plate 8a).

Euherdmania claviformis: dull translucent green, sand-covered; club-shaped but elongate individuals—a dense colony resembles packed sandy worm tubes; up to 1½ inches (3.8cm) tall. Entire coast, low intertidal and shallow subtidal.

Metandrocarpa (two species): dull to bright red; individuals 1/8- to ¼-inch (0.3-0.6cm) in diameter. *M. dura* (illustrated): individuals close-packed. Envelop seaweed stems; common in southern California and Baja California kelp beds at about 30-foot (9m) depth. *M. taylori*: individuals separate, resemble tiny mounds connected by thin sheet or stolons. Envelop sides or bottoms of rocks; entire coast, low intertidal to shallow subtidal.

Perophera annectans: pale green transparent matrix, loose to tightly packed clusters; individuals roughly spherical, diameter 1/8-inch (0.3cm). Envelop rocks and lower parts of algae; entire coast, low intertidal and shallow subtidal.

Pycnoclavella stanleyi: bright orange pharynx of animal protrudes from sandy tube about 1 inch (2.5cm) long, 1/16-inch (0.15cm) in diameter; many tightly packed individuals per colony. Northern and central California, low intertidal to about 20 feet (6m).

III. COMPOUND

This group of tunicates includes several genera and many species that can be distinguished only by painstaking dissection under a binocular microscope. Unfortunately, important and common animals are involved—the genera *Amaroucium, Eudistoma, Distaplia, Sigillinaria*, and *Synoicum*. These form thick layers or massive lobes of gelatinous material often enveloping undersides and walls of rocks, caves, cliffs, and ledges (see Plate 8b). A few compound tunicates can, however, be recognized on sight or by close inspection with a hand lens and can thus be treated here. The interested reader will have to identify the remainder in a suitably equipped laboratory with the aid of appropriate texts such as Light's Manual or the monograph on tunicates by W. G. Van Name.

Botrylloides diegense: matrix is dark, forming thin, irregular encrustation; tiny apertures surrounded by bright rings of green or orange give mottled appearance. Southern California harbors, shallow subtidal.

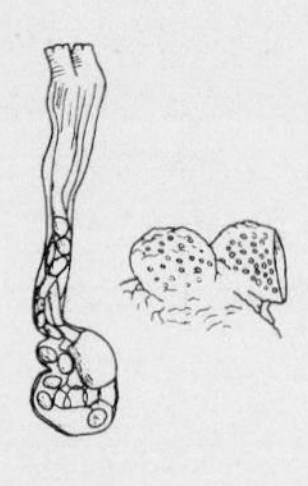

Cystodites (several species; *C. lobatus* [illustrated] common): translucent, pinkish to gray, gelatinous masses up to 1 inch (2.5cm) thick, 10 or more inches (25.4cm) across, often convoluted and lobed; matrix permeated with tiny discs, 1/16-inch (0.15cm) in diameter or less. Envelops rocks and algae; entire coast, low intertidal to 100 feet (30.5m) or more (Plate 8b).

Didemnum carnulentum: translucent, pink to white, forms thin, soft, smooth sheets a foot or more across. Closely resembles *Trididemnum* (below). Entire coast, intertidal to at least 80 feet (24.5m) (Plate 7c).

Metandrocarpa dura: red, envelops seaweed stems; described in group II (above) as a social tunicate.

Polyclinum planum: brown, olive, or light orange; opaque, soft, forms a flexible lobe or pad attached by thick stalk; up to 8 inches (20.3cm) tall. Entire coast, low intertidal to 80 feet (24.5m).

Synoicum sp.: translucent white to pink, often occurs as low mounds 1 or 2 inches (2.5 or 5.1cm) across, projecting through turf on rock tops; zooids visible to the eye as clear round areas within the colony matrix. Central California to Baja California, shallow subtidal to 100 feet (30.5m) (Plate 1a).

Trididemnum opacum: opaque, grayish white, forming thin, soft, smooth sheets spreading 1 foot (30.5cm) or more; can be peeled off. Resembles *Didemnum* (above). Entire coast, low intertidal to 100 feet (30.5m).

THE PELAGIC TUNICATES (SUBPHYLUM)

At times large numbers of transparent, jelly-like, cylindrical or barrel-shaped small animals abound in the nearshore waters or around the islands. They may occur singly or attached in chains up to 30 or 40 feet (9-12m) long and containing scores of individuals. Two genera of pelagic tunicates are common: one is a colonial form—*Pyrosoma*—while the other—*Salpa*—displays solitary individuals that give rise to chainlike aggregates by asexual budding. The solitary phase may differ significantly from the aggregate phase in appearances. Occasionally windrows of dead pelagic tunicates may accumulate on the bottom.

Pyrosoma giganteum: transparent to pink, luminesces in darkness; cylindrical, tapering slightly, hollow, roughened surface like a file, caused by protruding individuals of the colony; 1 inch (2.5cm) in diameter up to 2 feet (61cm) long. Generally far from shore, in oceanic waters.

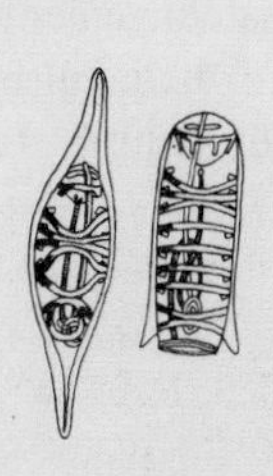

Salpa fusiformis: solitary form is transparent, flattened cylinder; 2 projections from one end, 9 translucent muscle bands, up to 3 inches (7.6cm) long. Aggregate form also transparent, elliptical, 6 muscle bands, 1 to 2 inches (2.5-5.1cm) long. Common in coastal and oceanic waters.

Salpa tilesii: solitary form transparent except for green organs, somewhat flattened, 2 projections from posterior end, 18-20 muscle bands; up to 8 inches (20.3cm) long. Aggregate form similar in coloration but no tail projections, 5 muscle bands; about 6 inches (15.2cm) long. Occasional in coastal and oceanic waters.

Chapter 12

MARINE ANIMALS: VERTEBRATES (FISHES AND MAMMALS)

FISHES

Fish mobility, often combined with shyness, can make underwater identification difficult. Behavior and the preferred location of a species (surface, midwater, bottom, sand, rock, crevices, etc.) are usually invaluable identification aids. General body shape is extremely important, but finer features such as fin outlines, spines, cirri, etc., may be concealed during the normal posture of an animal. Color may or may not be helpful.

Fishes are primarily divided into two groups: those with skeletons of cartilage (Condreicthys or Elasmobranch fishes—the sharks and rays) and fishes with bony skeletons (Osteichtheys or Teleost fishes). This classification is convenient here, since sharks and rays are relatively easy to distinguish. The bony fishes, however, are complex and require further subdivision.

CLASS ELASMOBRANCHI (SHARKS, RAYS)

SHARKS

Alopias vulpinus (Thresher): harmless; blue, brown, or black, huge upper tail lobe; 5 to 8 feet (1.5-2.5m) long. Entire coast, pelagic.

Carcharodon charcharias (Maneater, White Shark): dangerous; gray, tail lobes about equal, keel at base of tail; 15 feet (4.5m) long. Entire coast, pelagic and semi-benthic, shallow to at least 100 feet (30.5m).

Cephaloscyllium uter (Swell Shark): harmless; varies from yellow-gray to black, with white mottling; 3 feet (1m) long. Belly inflates with water when disturbed. Central California to Baja California, benthic and sluggish, to 100 feet (30.5m) or more.

Cetorhinus maximus (Basking Shark): harmless; grayish-dark above, white beneath, equal tail lobes, keel at base of tail; up to 30 feet (9m) long. Northern to southern California, pelagic.

Galeorhinus zyopterus (Soupfin): harmless; gray to blue above, light underneath; 6½ feet (2m) long. Often in schools. Entire coast, pelagic or semi-benthic, 5 to 1300 feet (1.5-396m).

Heterodontus (*Gyropleurodus*) *francisci* (Horn Shark): harmless; mottled, light gray to brown or black, spines ahead of dorsal fins, 4 feet (1.2m) long. Central California to Baja California, benthic and sluggish, intertidal to 500 feet (153m).

Isurus glaucus (Bonito Shark, Mako): dangerous; dark blue above, light underneath, tail lobes equal, keeled tail; up to 7 or 8 feet (2-2.5m) long. Central California to Baja California, pelagic.

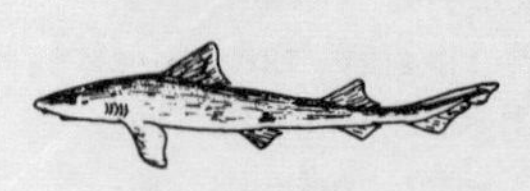

Mustelus californicus (Gray Smoothhound): harmless; dark gray, 3 or 4 feet (1-1.2m) long. Central California to Baja California, semi-benthic, shallow subtidal.

Prionace glauca (Blue Shark): dangerous; dark blue above, light underneath, large pectoral fins; length to 12 feet (4m). Entire coast, pelagic, occasionally semi-benthic.

Rhinotriacis henle (Brown Smoothhound): harmless; keel brown to bronze, 3 feet (1m) long. Entire coast, shallow and in bays.

Sphyrna zygaena (Hammerhead): dangerous; slate gray to brown, lateral projections from head contain eyes; 12 feet (4m) long. Southern California to Baja California, pelagic to semi-benthic, surface to at least 50 feet (15m).

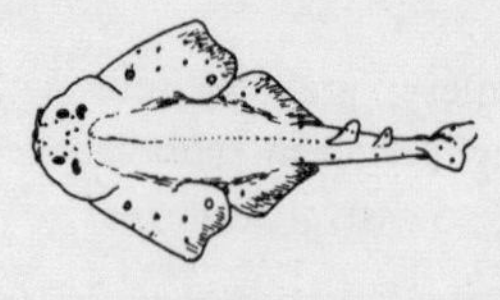

Squatina californica (Angel Shark): harmless; gray, brown, or black with mottling, flattened raylike body; 3 or more feet (1m) long. Entire coast, shallow subtidal to 200 feet (61m).

Triakis semifasciata (Leopard Shark): harmless; gray with black crossbars, 5 feet (1.5m) long. Entire coast, semi-benthic, shallow subtidal to about 50 feet (15m).

RAYS

All rays are benthic or semi-benthic.

Gymnura (*Pteroplatea*) *marmorata* (Butterfly Ray): brown with light spots, diamond-shaped, small or no spine in short tail; length to 4½ feet (1.5m). Mainly in bays, occasionally open coast, southern California to Baja California, shallow to 30 feet (9m).

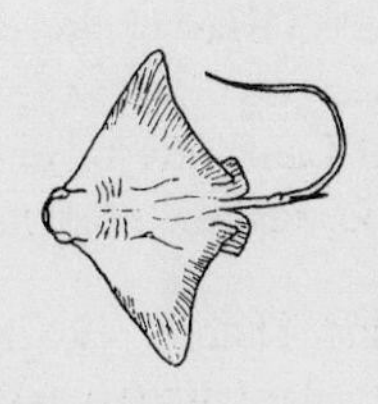

Holorhinus (*Aetobatus*) *californicus* (Bat Ray): olive to black above, light underneath, diamond-shaped, long tail with one or more large spines; 4 feet (1.2m) across. Entire coast, shallow subtidal to 150 feet (46m).

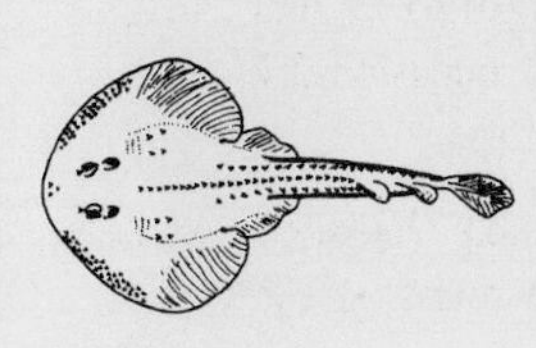

Platyrhinoides triseriata (Thornback Ray): silver to brown above, white underneath, 3 rows of small thornlike spines dorsally; 3 feet (1m) long. Central California to Baja California, shallow to 100 or more feet (30.5m).

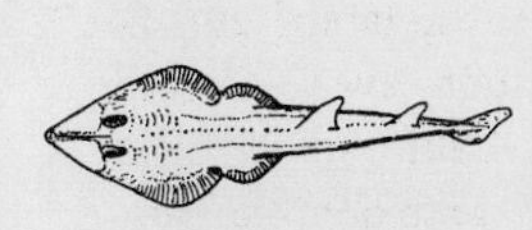

Rhinobatos productus (Shovelnose Guitarfish): gray to brown above, light underneath, elongated diamond-shaped; 4 feet (1.2m) long. Central California to Baja California, shallow to 50 feet (15m).

Squatina californica (Angel Shark): discussed under Sharks (above).

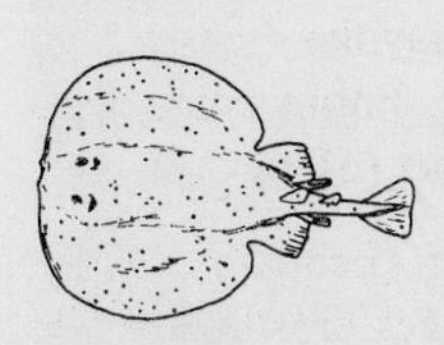

Torpedo (*Tetranarce*) *californica* (Electric Ray): dangerous, gives powerful electric shock; grayish-blue with dark spots, round; 3 feet (1m) across. Entire coast, 30 to 200 feet (9-61m).

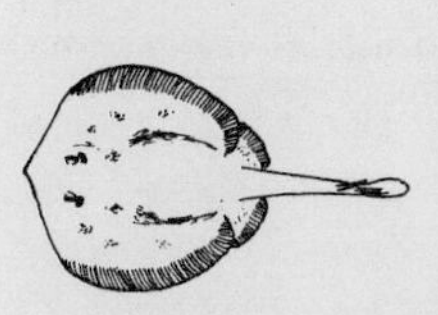

Urobatis halleri (Round Stingray): gray to brown above, white underneath, tail with spine; 20 inches (50.8cm) long. Southern California to Baja California, shallow to 70 feet (21.5m).

Zapteryx exasperatus (Mottled Guitarfish): gray with dark blotching above, light underneath, roughly diamond-shaped; 4 feet (1.2m) long. Southern California to Baja California, shallow to 70 feet (21.5m).

CLASS OSTEICHTHYES (BONY FISHES)

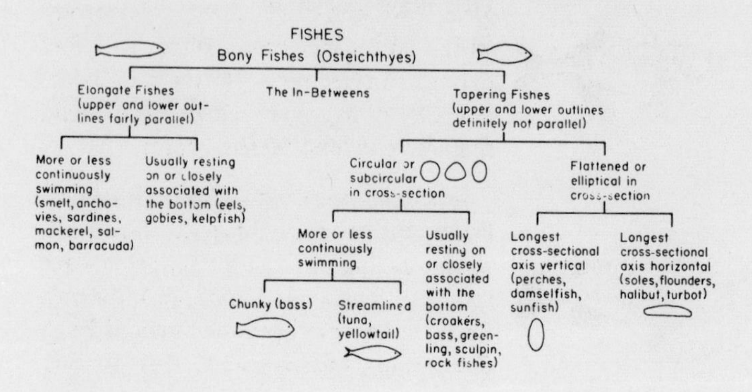

SWIMMING BASSLIKE FISHES

Myctoperca jordani (Gulf Grouper) (illustrated): olive-gray to black above, light olive beneath, sometimes with 4 rows of dark blotches, projecting lower jaw, straight tail (distinguishes from broomtail grouper), to 3½ feet (1m) long. *M. xenarchus* (Broomtail grouper): similar to gulf grouper, except may have a blue hue, is usually spotted, and tail is scalloped; length to 5 feet (1.5m). Southern California and entire Baja California coast, shallow to at least 60 feet (18m).

Paralabrax clathratus (Kelp Bass, Calico Bass): brown to olive-yellow above, yellow to dirty white beneath, mottled with light patches; length to 26 inches (66cm). Central California to Baja California, surface to 130 feet (40m). (*P. nebulifer* and *P. maculofasciata* are discussed under Semi-Benthic Basslike Fishes, below.)

Roccus saxatilis (Striped Bass): brassy-green to bluish-black above, sides silvery with 7 to 8 dark stripes, length to 2 feet (61cm). Anadromous. Northern to southern California, rare below Morro Bay; shallow to at least 50 feet (15m).

Sebastes mystinus (Blue Rockfish, Priest Fish): dark blue to gray-black above, fading to light belly, often dark spotting; length to 20 inches (50.8cm). Entire coast, very common north of Point Conception; surface to 130 feet (40m) in cold water.

Stereolepis gigas (Black Sea Bass, Jewfish): bluish or dark brown to black above, fading to lighter on lower sides and belly, sometimes blotched; length to 7½ feet (2.5m). Distinguish from groupers (above) by blunt nose, lower jaw doesn't protrude. Central California to Baja California, surface to more than 100 feet (30.5m).

SEMI-BENTHIC BASSLIKE FISHES

Cheilotrema saturnum (Black Croaker): dusky to black above, fading to silvery belly, pale vertical band mid-body, black spot on gill cover; length to 16 inches (40cm). On sand or rocks near sand, southern California and Baja California, surface to about 100 feet (30.5m).

Genyonemus lineatus (White Croaker): dark silver above with brassy luster, fading to silver sides and belly, faint wavy diagonal striping, black spot at base of pectoral; length 13 inches (32.5cm). On sandy bottom in bays and open sea, entire coast, shallow to 30 feet (9m).

Hexagrammos decagrammus (Greenling, Seatrout): male is green, brown, or copper, sparsely spotted; female is light brown, densely spotted; length to 21 inches (52.5cm). Northern and central California, shallow to at least 50 feet (15m).

Menticirrhus undulatus (Corbina): discussed under Semi-Benthic Elongate Fishes (below).

Oxylebius pictus (Convict Fish): gray to light or reddish brown, mottled with 6 to 7 vertical bars; length to 12 inches (30cm). Entire coast, surface to 130 feet (40m).

Paralabrax maculofasciata (Spotted Bass): gray to greenish-brown above, shading to light olive below, dense reddish spotting; length to 18 inches (45cm). Southern California and Baja California, 10 to at least 30 feet (3-9m).

Paralabrax nebulifer (Sand Bass): gray to greenish above, fading to gray or white beneath, dusky banding or splotches, spotting on head; length to 22 inches (55cm). On sand or rock, central California to Baja California, shallow to 130 feet (40m).

Roncador stearnsi (Spotfin Croaker): brassy-green to gray above, fading to silvery belly, large black spot at base of pectoral fin; length to 2 feet (61cm). On sandy bottom in bays and open sea, southern California and Baja California, shallow to 50 feet (15m).

Scorpaena guttata (Sculpin, Scorpionfish): tan to reddish-brown above, pinkish underneath, mottled and spotted with red, brown, purple; length to 18 inches (45.7cm). Resembles a stubby cabezone (below). (Beware of venomous spines!) Entire coast, intertidal to 130 feet (40m) or more.

Scorpaenichthys marmoratus (Cabezone): tan, green, brown, or red with blotches or mottling; length to 30 inches (76cm). Lack of scales and elongate body distinguish from sculpin (above). (Do not eat poisonous roe; otherwise good food.) Central California to Baja California, shallow to 130 feet (40m).

Sebastes (Rockfishes; many species): lower jaw strongly projecting or forming leading edge of pointed snout; large eye, forehead usually strongly sloping, lower jaw usually flat; edge of upper gill flap forms a sharp angle, often with spines; strong prominent spines in front part of dorsal and anal fins. *S. atrovirens* (Kelp Rockfish) (illustrated): light brown with olive-brown mottling, creamy underneath with spotting, upper gill flap black. *S. carnatus* (Gopher Rockfish) and *S. chrysomelas* (Black-and-tan Rockfish): dark brown background with light blotches, yellow underneath; blotches are flesh-colored in *S. carnatus*, yellow in *S. chrysomelas*, but usually not distinguishable underwater (Plate 2d). *S. miniatus* (Vermilion Rockfish): vermilion above, shading to pink below, blotches formed of clusters of black dots, 3 faint orange stripes radiate from eye. *S. mystinus* (Blue Rockfish): described under swimming Basslike Fishes (above). *S. serranoides* (Olive or Bass Rockfish) and *S. flavidus* (Yellowtail Rockfish): gray to olive-brown above, fading to lighter below, light splotches on back. *S. serriceps* (Treefish): 6 to 7 dark and yellow alternating bands, conspicuous red lips. All these rockfishes are relatively small (14 to 18 inches [35.6-45.7cm] maximum). All occur along entire coast in cold water (except

S. serranoides, which only goes south as far as San Francisco); all range to 130 feet (40m) or deeper.

Umbrina roncador (Yellowfin Croaker): brassy-gray or green above, shading to silver below, diagonal wavy blue-green lines, lacks black spot of spotfin croakers; length to 16 inches (40.6cm). Over sand in surf or bays, southern California and Baja California, usually shallow to 25 feet (8m).

SWIMMING PERCHLIKE FISHES

Amphistichus argenteus (Barred Surfperch): olive, blue, or gray on back, extending down sides as bars alternating with columns of spots, interspersed with silver extending to belly; length to 15 inches (38.1cm). Over sand in bays and open sea; entire coast, surface to 30 feet (9m).

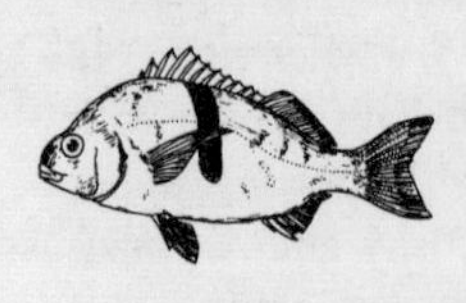

Anisotremus davidsoni (Sargo): silvery-gray back, silver sides and belly, prominent dark bar, tail slightly notched (distinguishes from juvenile *Rhacochilus vacca*, below); length to 14 inches (35.6cm). In bays and open sea, southern California and Baja California, surface to 130 feet (40m).

Brachyistius frenatus (Kelp Perch): olive-brown back, coppery-silver sides and belly, snout turns up (distinguishes from *Cymatogaster*, below); length to 8 inches (20.3cm). Entire coast, surface to 100 feet (30.5m).

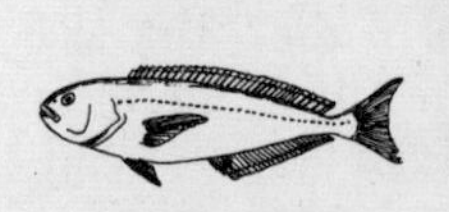

Caulolatilus princeps (Ocean Whitefish): yellow-brown back, fading to whitish below, fins tinged with blue, yellow, and green; dorsal and anal fins form a continuous line almost to tail; length to approximately 40 inches (1m). Central California to Baja California, surface to 130 feet (40m).

Chromis punctipinnis (Blacksmith): dark blue back, fading to slate blue below, small spots along back and tail, turned-up snout; length to 10 inches (25.4cm). Southern California and Baja California, surface to 150 feet (46m).

Cymatogaster aggregata (Shiner): green-silvery back, bright silver below, faint dusky horizontal stripes and yellow vertical bands; straight silhouette above eye (distinguishes from *Brachyistius*, above); length to 7 inches (17.8cm). Entire coast, surface to 30 feet (9m).

Embiotoca jacksoni (Black Perch): variable color–brown commonest, also red, tan, green, blue–fading below, usually conspicuous vertical banding; length to 14 inches (35.6cm). Common, entire coast, surface to 130 feet (40m).

Embiotoca lateralis (Striped Perch): dark green back, usually fading below, often alternating blue and red fin stripes, tail often yellow, coppery bar above nose; length to 14 inches (35.6cm). Entire coast, surface to 60 feet (18m).

Girella nigricans (Opaleye): dark or olive back, green sides, lighter belly, frequently one or more white spots centrally on back, eye opalescent blue; length to nearly 20 inches (50.8cm). Entire coast, surface to 100 feet (30.5m).

Hyperprosopon argenteum (Walleye): blue back, silvery white sides and belly; length to 9 inches (23cm). Black-tipped pelvic fins and deep body distinguish from *Phanerodon* (below). Over sandy bottom; entire coast, surface to 40 feet (12m).

Hypsurus caryi (Rainbow Perch): horizontally striped in red, orange, and blue, vertical orange bars, silhouette straight underneath; length to 10 inches (25.4cm). Entire coast, surface to 130 feet (40m).

Hypsypops rubicunda (Garibaldi): brilliant orange, juveniles streaked and spotted with blue; length to 15 inches (38cm). Southern California and Baja California, surface to 100 feet (30.5m) (Plate 4e and Cover).

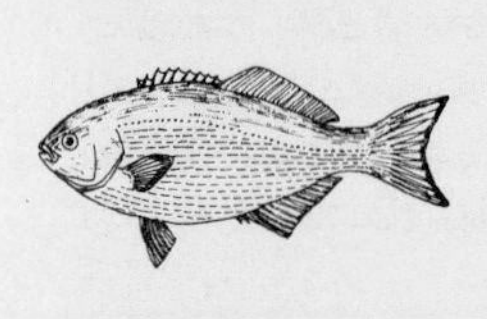

Medialuna californiensis (Halfmoon): dark slate-blue back, fading to whitish below, sometimes faint mottling on sides; tail often appears strongly forked; length to 12 inches (25.4cm). Central California to Baja California, surface to 130 feet (40m).

Micrometrus minimus (Dwarf Perch): greenish back, splotchy sides, silvery beneath, horizontal striping on lower sides; length to 3 inches (7.6cm). Tidepools, open sea, and bays, central California to Baja California, intertidal to 30 feet (9m).

Mola mola (Sunfish): dark blue above, light blue to silvery beneath, tail rounded; length to 10 feet (3m). Entire coast, pelagic in open sea.

Palometa simillima (Pampano): green or blue back, fading to silvery beneath, tail deeply notched (distinguishes from *Caulolatilus*, above); length to 11 inches (28cm). In bays and open coast over sand, entire coast, surface to 30 feet (9m).

Phanerodon furcatus (White Perch): dusky back, white silvery sides and belly, pelvic fin clear (distinguishes from *Hyperprosopon*, above); length to 12 inches (30.5cm). In bays and open sea, entire coast, surface to 130 feet (40m).

Rhacochilus toxotes (Rubberlip): back gray to dark blue, fading to light gray or white beneath, pectoral fins yellow, thick white to pink lips; length to 18 inches (45.7cm). Juveniles may have one or two vertical bars. Entire coast, surface to 130 feet (40m).

Rhacochilus vacca (Pile Perch): back olive, gray, brown, or black, fading to silvery on lower sides and belly, juveniles usually silvery with vertical dark bars, tail deeply notched (distinguishes from *Anisotremus*, above); length to 16 inches (40.6cm). Entire coast, surface to 130 feet (40m).

FLATFISHES

All flatfishes have whitish bellies; color of upper surface often changes temporarily to match surroundings.

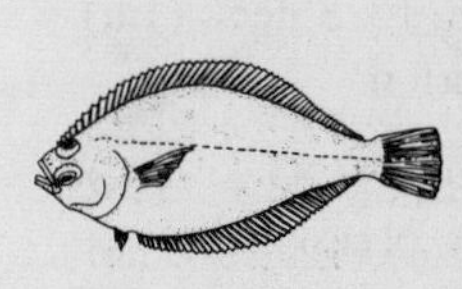

Citharicthys (Sand Dabs): light tan to olive, speckled, lateral lines almost straight. *C. sordidus* (Pacific Sand Dab) (illustrated): lower eye larger than snout, length to 12 inches (30.5cm). *C. stigmaeus* (Speckled Sand Dab): lower eye smaller than snout, length to 8 inches (20.3cm). Entire coast, 10 to about 300 feet (3-91m).

Eopsetta jordani (Petrale Sole): olive to brown, slightly curving lateral line, no ridge between eyes; length to 20 inches (51 cm). Entire coast, but rare south of Monterey.

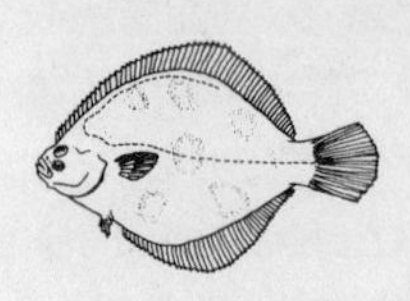

Hypsopsetta guttulata (Diamond Turbot): dark greenish-brown to brown, very broad, no ridge between eyes, recurved lateral line; length to 18 inches (45.7cm). Entire coast, shallow to 140 feet (43m).

Microstomus pacificus (Dover Sole): yellow to brown, blotched, fins tipped with black, elongate elliptical body; length to 24 inches (61cm). Entire coast, but rare south of Monterey.

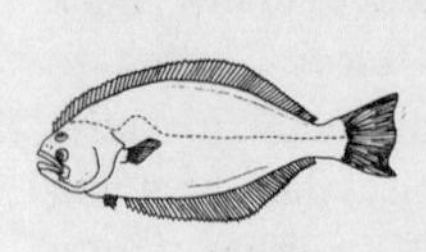

Paralicthys californicus (California Halibut): gray to greenish-brown, sometimes mottled, elongate body, jaw extends backward to or beyond eyes; length to 3½ feet (1m). Central California to Baja California, shallow to at least 100 feet (30.5m).

Parophrys vetulus (English Sole): various shades of brown, ridge between eyes, recurving lateral line, narrow body, pointed head; length to 21 inches (53.3cm). Entire coast, but rare south of Santa Barbara.

Platicthys stellatus (Starry Flounder): dark brown with vague blotching, black, white, and orange stripes on fins; lateral line approximately straight; length to 3 feet (1m). Entire coast, but rare south of Point Conception.

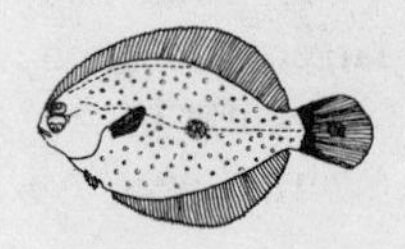

Pleuronichthys coenosus (C-O Turbot): dark brown, black, or occasionally variegated color, banding on tail usually resembles the letters CO; ridge between eyes; length to 14 inches (35.6cm). Entire coast, shallow to 100 feet (30.5m).

Pleuronichthys decurrens (Curlfin Turbot): light to dark brown with blotching, ridge between eyes, recurving lateral line with little arching over pectoral fin; length to 12 inches (30.5cm). Northern to southern California.

Xystreurys liolepis (Fantail Sole): olive to brown, blotched, eyelike spot behind head and sometimes another nearer tail; arching, lateral line; length to 15 inches (38cm). Central California to Baja California, 30 to at least 100 feet (9-30.5m).

SWIMMING ELONGATE FISHES

Atherinops affinis (Topsmelt): blue-gray to green back, bright silver stripe bordered above in blue along sides, silvery below; slightly overlapping upper jaw, 2 dorsal fins, long pectoral fin; length to 12 inches (30.5cm). Entire coast, surface to 80 feet (24m).

Atherinopsis californiensis (Jacksmelt): gray-green to blue-green, metallic blue stripe along sides, silvery below; jaws equal, 2 dorsal fins, small pectoral fin; length to 18 inches (45.7cm). Entire coast, surface to about 100 feet (30.5m).

Cololabis saira (Saury): deep blue to dark green back, silver below; pointed jaws, lower jaw protruding; length to 14 inches (35.6cm). Entire coast, generally several miles (km) offshore.

Cynoscion nobilis (White Seabass): blue to blue-gray back, silvery below, blunt snout; length to 4 feet (1.2m) but mostly less than 2 feet (61cm). Juveniles with 3 to 6 vertical bars. Entire coast, surface to 50 feet (15m).

Cypselurus californicus (Flying Fish): deep blue back and sides, silvery belly, very long pectorals; length to 18 inches (45.7cm). Southern California and Baja California, pelagic in open sea.

Engraulis mordax (Northern Anchovy): metallic blue or green back, silvery below; greatly overlapping upper jaw, single dorsal fin; length to 8 inches (20.3cm). Entire coast, pelagic.

Leuresthes tenuis (Grunion): blue-green to gray-green back, silver stripe bordered above by violet along both sides; 2 dorsal fins, no teeth; length to 7 inches (18cm). Central California to Baja California, shallow from intertidal to 1 mile (1.6km) offshore.

Oncorhynchus tschawtscha (King Salmon): blue to dark gray back with dark spots extending to tail, silver sides and belly; upper outline fairly straight, lower outline curved; length to almost 5 feet (1.5m), usually less than 2 feet (61cm). Anadromous. Northern and central California.

Oxyjulis californicus (Senorita): yellow-brown back, lighter below, head with brown and blue streaks, black band on tail; length to 8 inches (20.3cm). Common, central California to Baja California, surface to 130 feet (40m).

Pneumatophorus japonicus (Pacific Mackerel): dark green to blue back, fading to iridescent silver below, about 30 wavy streaks diagonally across back; length to 25 inches (63.5cm). Entire coast, surface to at least 100 feet (30.5m).

Salmo gairdneri (Steelhead Trout): steel-blue back, densely spotted, silvery sides and belly; top outline straight, belly curving; length to 33 inches (84cm).

Anadromous. Entire coast, but rare south of Monterey.

Sardinops caerula (Sardine): dark green to blue back, darkly spotted, shading to silvery sides and belly, one or more rows of black spots (laterally); single dorsal fin, rounded snout, prominent lower jaw; length to 14 inches (35.6cm). Entire coast, surface to at least 40 feet (12m).

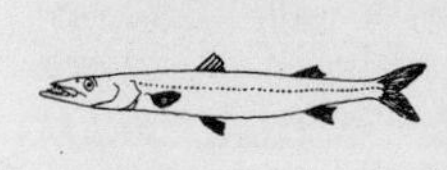

Sphyraena argentea (Barracuda): dark gray or brownish back, silvery below, yellowish tail, pointed snout with projecting lower jaw; length to about 4 feet (1.2m). Entire coast, usually near surface but sometimes down to 40 feet (12m) or more.

Trachurus symmetricus (Jack Mackerel): green with blue iridescence on back, often mottled, silver below; projecting lower jaw, conspicuous lateral line posteriorly; length to 22 inches (56cm). Entire coast, surface to 130 feet (40m).

TUNALIKE FISHES

Katsuwonus pelamis (Skipjack): blue back, silvery below, 4 or 5 horizontal stripes on lower sides; length rarely exceeds 2 feet (61cm). In open sea, southern California and Baja California; rare farther north.

Neothunnus macropterus (Yellowfin Tuna): dark blue above, fading to gray-silver below, yellow band along sides, very long pectoral fins; length to at least 5 feet (1.5m). Southern California and Baja California, pelagic.

Sarda lineolata (Bonito): blue to violet back with 10 to 11 black diagonal stripes, shading to silvery below; length to 40 inches (1m). Entire coast, well out to sea.

Seriola dorsalis (Yellowtail): blue to green above, ending at brassy stripe horizontally along sides, silver below, tail greenish-yellow; length to 5 feet (1.5m) but rarely exceeds 3 feet (1m). On outer edges of kelp and beyond; southern California and Baja California, surface to 50 feet (15m) or more.

Thunnus saliens (Bluefin Tuna) (illustrated): deep iridescent blue back, silver below, irregular white spotting, short pectoral fin (distinguishes from Yellowfin Tuna, above, and Albacore); length to about 30 inches (76cm). *T. germo* (Albacore): similar, but with very long pectoral fin. Entire coast, but rare north of Point Conception; occurs well offshore.

SEMI-BENTHIC ELONGATE FISHES

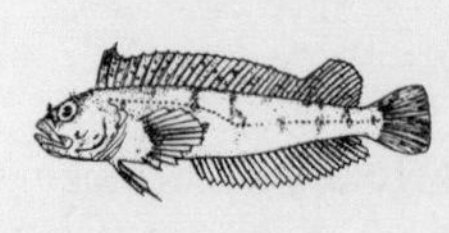

Alloclinus holderi (Island Kelpfish): light gray, 7 faint cross bands on[sdes; length to 4 inches (10.2cm). Southern California and Baja California, intertidal to 70 feet (21m).

Cebidicthys violaceus (Monkeyface Eel): dull to brownish-green above, fading below (single lateral line distinguishes from *Xiphister*, below); length to 30 inches (76cm). Northern and central California, rarely south; intertidal to about 10 feet (3m), usually under rocks.

Coryphopterus nicholsii (Bluespot Goby): pale semi-transparent body, prominent dark eyes; length to 5 inches (12cm). Entire coast, intertidal to 2100 feet (640m).

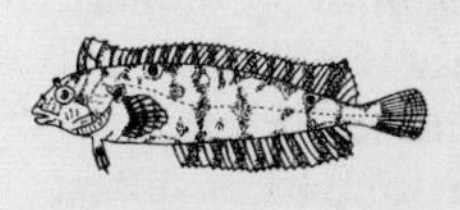

Gibbonsia elegans (Spotted Kelpfish): variable from green, brown, red, to lavender, spots along sides; length to 4 inches (10.2cm). Entire coast, intertidal to 70 feet (21m).

Gibbonsia metzii (Weed Kelpfish): similar to *G. elegans* (above) but lacks spotting; length to 5 inches (12.7cm). Entire coast, shallow to 30 feet (9m).

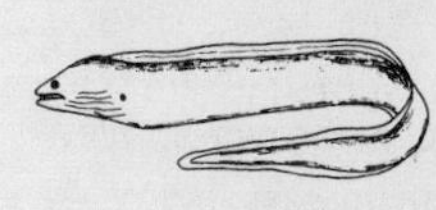

Gymnothorax mordax (Moray Eel): brown with white mottling, tapering snout; length to 5 feet (1.5m). Southern California to Baja California, intertidal to 130 feet (40m) (Plate 2).

Heterostichus rostratus (Giant Kelpfish): variable–green, purple, brown, or black–with blotching, lighter below; pointed snout; length to 16 inches (40.6cm). Entire coast, shallow to 100 feet (30.5m).

Hypsoblennius gilberti (Tidepool Blenny): variable, tan to dark olive, mottled, maximum diameter at head; length to 5 inches (12.7cm). Southern California to Baja California, intertidal to shallow subtidal.

Lythripnus (Goby): coral red with brilliant blue striping. *L. dalli* (Catalina Goby) (illustrated): 4 to 6 vertical blue bands; length to 1¼ inches (3.2cm). Southern California and Baja California, intertidal to 300 feet (91m). *L.* (*Zonogobius*) *zebra* (Zebra Goby): 15 bands, length to 1¾ inches (4.5cm). Southern California and Baja California, intertidal to 100 feet (30.5m) or more.

Menticirrhus undulatus (Corbina): dusky to dark blue back, white sides and belly; length to 20 inches (51cm). On sand in surf, southern California and Baja California, 2 to 30 feet (0.6-9m) deep.

Neoclinus blanchardi (Sarcastic Fringehead): gray to brown or red, mottled, long branching cirri (fringes) on forehead; length to 8 inches (20.3cm). Central and southern California, 25 to 90 feet (8-27m).

Ophiodon elongatus (Ling Cod): variable–green, blue, black, or brown above, gray to white with bluish cast below–blotched; length to 5 feet (1.5m). Entire coast, shallow to 130 feet (40m).

Paraclinus (*Auchenopterus*) *integripinnis* (Reef Finspot): brown with mottling, base of pectoral fin violet with black border, black spot on hind dorsal fin; length to 2½ inches (6.4cm). Southern California to Baja California, shallow.

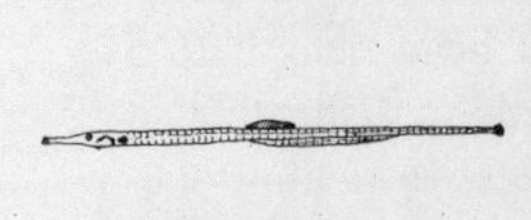

Syngnathus (Pipefish): tubelike body, entwines itself in vegetation. *S. californiensis* (Great Pipefish): olive, brown, or red with speckling; length to 18 inches (45.7cm). On open coasts, shallow to 50 feet (15m). *S. leptorhyncus* (Lesser Pipefish) (illustrated): brown to green, length to 6 inches (15.2cm). In eel grass of bays, southern California and Baja California; intertidal to 20 feet (6m).

Xiphister mucosus (Rock Eel): greenish-black with mottling, somewhat paler below, white to brown streaks radiate from eye, 4 lateral lines (distinguishes from *Cebidichthys*, above); length to 2 feet (61cm). Northern and central California, uncommon farther south, intertidal to shallow subtidal, usually under rocks.

THE IN-BETWEENS (Fishes Not Falling into Previous Categories)

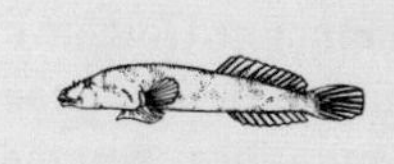

Gobiesox (*Arbaciosa*) *rhessodon* (Clingfish): pale olive to dark, belly white, encircled midway by white band, sucker beneath head; length to 2½ inches (6.4cm). Southern California to Baja California, intertidal to 60 feet (18m).

Halichoeres semicinctus (Rock Wrasse): dark brown to greenish, males have blue band behind head, black band above pelvic fin; length to 14 inches (35.6cm). Southern California to Baja California, shallow to 60 feet (18m).

Pimelometopon pulchrum (Sheephead): male has black head and tail, white chin, encircled midway by broad pink to red band; female has pink to reddish back and sides, whtie chin and belly; length to 3 feet (1m) (males). Central California to Baja California, shallow to 150 feet (46m).

Seriphus politus (Queenfish): bluish to bronze back, fading to silvery below, yellowish fins; length to 12 inches (30.5cm). On sandy bottoms, central California to Baja California, surface to 40 feet (12m).

THE MARINE MAMMALS (CLASS MAMMALIA)

California marine mammals range from the 50-pound sea otter to the 30-ton gray whale. None are aggressive or known to be dangerous. Most are quite shy and are encountered only briefly and by accident. Nonetheless, they are capable of harming a diver and should never be provoked by jabbing or hitting. Harbor seals and sea lions at times are actually playful. They have been known, however, to take speared fish away from a diver forcibly.

Delphinus delphis (Common Dolphin): black above with white and light gray markings along head and sides, pointed beak; length to 7 feet (2.2m). Common, entire coast, pelagic.

Enhydra lutris (Sea Otter): brown (tan when old), rotund body; length 3 to 5 feet (1-1.5m). Uses forepaws to break shellfish open against rocks held on chest. Central California, surface to 50 feet (15m).

Eschrictius gibbosus (*glaucus*) (Gray Whale): dark to light gray back, knobby outline posteriorly, no dorsal fin; length 30 to 50 feet (9-15m). Entire coast, surface to 100 feet (30.5m).

Eumetopias jubata (Steller Sea Lion): gray to brown, whitish blotching; length to 13 feet (4m). Northern and central California, intertidal to 100 feet (30.5m) or more.

Globicephala scammoni (Pilot Whale, Blackfish): black above, light patch behind dorsal fin, spherical forehead; length to about 20 feet (6m). Entire coast, surface to more than 100 feet (30.5m).

Lagenorhynchus obliquidens (Striped Dolphin, White-sided Dolphin): black back, light underneath, light stripes on posterior back, dorsal fin often with light patch, short beak; length up to 7 feet (2.2m). Common in large schools, entire coast.

Mirounga angustirostris (Northern Elephant Seal): light to dark brown, adult males have elongate overhanging proboscis; length to about 15 feet (4.5m). Entire coast, intertidal to at least 50 feet (15m).

Orcinus orca (Killer Whale): black with white markings, white spot behind eye, males have high triangular dorsal fin; length to 25 feet (7.5m). Entire coast, surface to at least 50 feet (15m).

Phoca vitulina (Harbor Seal): gray to black, with white spotting; length to 5 feet (1.5m). Entire coast, intertidal to at least 60 feet (18m).

Phocoena vomerina (Bay Porpoise): dark gray above, lighter below; length to about 6 feet (2m). Bays, rivers, and inshore waters, northern and central California.

Phocoenoides dalli (Dall's Porpoise): black with brilliant white markings, small flippers; length to 6 feet (2m). Northern and central California, occasionally farther south; pelagic.

Tursiops truncatus (Bottlenose Dolphin): grayish with lighter markings, beak reduced, mouth curves into a smile; length to 12 feet (4m). Southern California and Baja California, pelagic.

Zalophus californianus (California Sea Lion): tan to brown, adult males have crested forehead; length to about 7 feet (2.2m). Common, entire coast, intertidal to more than 100 feet (30.5m) (Plate 8c).

GLOSSARY

anadromous–moving from salt water to fresh in order to reproduce
anastomosing–coming together
Anterior–front or forward region of an animal
aperture–opening
apex–region of a shell where a point is formed
arborescent–treelike
avicularia–beaklike appendage of a bryozoan
axial–associated with an axis
axis–imaginary line through the body of an animal which prescribes symmetry

beak (of shell)–point or apex of a shell
benthic–referring to organisms living on or closely associated with the ocean bottom
bifurcated–divided
biota–living organisms present at a site
blade–leaf-like portion of a macroalga

calcareous–composed of calcium carbonate
calycophoran–refers to a suborder of the siphonophores
canopy–portion of large macroalga which spreads across the ocean surface
canyons, submarine–underwater valleys caused by erosion
carapace–hard chitinous exterior shell covering the thorax
chelae–pincers
cilia–microscopic fields of filaments
ciliary–of or referring to cilia
commensal–association of two animals living together in which one animal benefits while neither harming nor benefiting the other
continental shelf–gradually sloping region extending from the shoreline to a depth of 100 or 200 meters
convoluted–deeply folded
crenulation–a fold
current–definable water movement

deck (of shell)–transverse plate of shelly material
deep littoral–ocean region extending over the continental shelf from 50 to 200 feet
detritus–non-living microscopic debris suspended in the water or accumulated on the bottom
disk (anemone)–flat region of the sea anemone which is surrounded by tentacles
dorsal–animal surface that is oriented upward

eudoxome–free-living stage of a siphonophore

fasciole–spiral band around a gastropod shell
foot (of snail etc.)–large muscular organ sometimes used for locomotion or burrowing
forcipate–resembling forceps

girdle–tough integument surrounding valve of mollusk
globose–globular
gonad–reproductive organ

intertidal–ocean bottom over which the high and low tides range

keel–outstanding rib marking abrupt change in the slope of a shell

lamina–flat blade of a macroalga
lateral–of or relating to the side
ligament–band of tough elastic fibers uniting the two valves of a bivalve
lobes–rounded projections for extensions of the body
lunule–heart-shaped area in front of the beaks of a bivalve

mantle–fold of epidermis which secretes the shells of some marine animals

neritic province–water covering the continental shelf
nodule–small, beadlike protuberances on chiton valves

oceanic province–waters outside the continental shelf
operculum–shell or hard material used to seal off an aperture
oscule–pore or small opening

papillae–tiny fingerlike projections
pelagic–referring to organisms suspended and living in the water
periostracum–brown protein shell covering
plush–plentiful
pneumatocyst–glas bladder of a macroalga
polyp–basic body shape of many lower animals; e.g. corals, bryozoans
posterior–back region

quadrate–four-sided

radial–circular
recurved–curved back upon itself
reticulated–networklike structure
rostrum–beaklike process

sail–flat, vertical organ acted on by wind in some animals
salinity–salt content of the water
scuba zone–ocean waters accessible to scuba divers, above 200 feet in depth
sediment–deposits of fine-grained materials
septa–vertical walls within an internal body cavity
sessile–stationary
shallow littoral–ocean waters covering the continental shelf 30 to 50 feet in depth
siltation–accumulation of fine-grained sediments
spicule–needlelike skeletal structure
spire (of shell)–coiled apex of snail shell
spire angle–angle subtended by the apex of a snail shell
stalk (in animals)–stemlike organ attaching an animal to the substrate
stipe–vinelike portion of a macroalga
submarine canyon–underwater valley caused by erosion
surge–strong current of short duration
suture–line delimiting region between old and newly formed shell

telson–tail appendages in crustaceans
tentacles–elongated flexible protuberances often used in food gathering
test–calcareous internal shell
theca–hard external shell of hydroids
thermocline–distinct temperature discontinuity
transverse–direction perpendicular to the longest axis
trigonal–three-sided
tubercle–small projection
tuberculated–covered with small projections
turbidity–cloudiness or murkiness of the water
turbulence–random, swirling water movements

umbilicus–depression in the central axis
umbrella–hemispherical portion of the jellyfish medusa
upwelling–vertical currents which bring deep water to the surface

valve–shell, often referring to that of a mollusk
varices–elevated outer lip of certain gastropod shells indicating periodic resting stages
ventral–animal surface that is oriented downward

zonation–formation of zones, often denoting the vertical limit of organisms
zooecium–skeletonized envelope around an individual bryozoan

REFERENCES

Abbott, R. Tucker, 2d ed., 1974.
American Seashells. Van Nostrand, N.Y., 541 pp.

Abbott, R. Tucker, and Herbert S. Zim, 1968.
A guide to field identification: Seashells of North America. Golden Press, N.Y., 280 pp.

Anonymous, 1971.
Biology of Giant Kelp beds (*Macrocystis*) in California (ed. by W. J. North). J. Cramer, Lehre, Germany, 600 pp.

Barnhart, Percy Spencer, 1936.
Marine Fishes of Southern California. Univ. of Calif. Press; 209 pp.

Boolootian, Richard A., and David L. Leighton, 1966.
A Key to the Species of Ophiuroidea (Brittle Stars) of the Santa Monica Bay and Adjacent Areas. Los Angeles County Museum of Science, No. 93, pp. 1-20.

Cannon, Ray, 1964.
How to Fish the Pacific Coast. Lane Book Co., Menlo Park, Calif.; 337 pp.

Carlisle, John G., Charles H. Turner, and Earl E. Ebert, 1964.
Artificial habitat in the marine environment. Calif. Dept. of Fish and Game, Fish Bul. 124; 93 pp.

Church, Jim, and Cathy Church, 1972.
Beginning underwater photography. Pub. by the authors from articles in Skin Diver Magazine (see under List of Journals).

Cox, Keith W., 1960.
Review of the abalone in California. Calif. Fish and Game, Vol. 46, no. 4, pp. 381-406.

Daugherty, Anita E., 1966.
Marine mammals of California. Calif. Dept. Fish and Game, Sacramento, 87 pp.

Dawson, E. Yale, 1966.
Seashore Plants of Northern California. Univ. Calif. Press, Berkeley, Calif. Nat. Hist. Guide 20, 103 pp.

Dawson, E. Yale, 1966.
Seashore plants of Southern California. Univ. Calif. Press, Berkeley, Calif. Nat. Hist. Guide 19, 101 pp.

Dawson, E. Yale, Michael Neushul, and Robert D. Wildman, 1960a.
Seaweeds associated with kelp beds along southern California and northwestern Mexico. Pacific Naturalist, Vol. 1, No. 14, 81 pp.

DeLaubenfells, Maurice W., 1932.
The Marine and Freshwater Sponges of Calif. Proc. U.S. Nat. Mus., Vol. 81, pp. 1-140.

Emery, K. O., 1961.
The Sea Off Southern California: A Modern habitat of Petroleum. John Wiley, N.Y., 366 pp.

Fager, Edward W., 1963.
Communities of organisms. The Sea, ed. M. N. Hill. J. Wiley and Sons, New York and London, Vol. 2, pp. 415-437.

Fitch, John E., 1953.
Common Marine Bivalves of California. Dept. Fish and Game, Sacramento, Fish Bull. 90, 102 pp.

Fitch, John E., and Robert J. Lavenberg, 1971.
Marine food and game fishes of California. Univ. Calif. Press, 179 pp.

Frey, Hank, and Paul Tzimoulis, 1968.
Camera below. Association Press, 291 Broadway, New York, N.Y. 10007.

Furlong, Marjorie and Virginia Pill (text), Ellis H. Robinson (photos), 1970.
Starfish, methods of preserving and guides to identification. Ellison Ind., Edmonds, Wash.

Hartman, Olga, 1968.
Atlas of the Errantiate Polychaetous Annelids from California. Univ. of So. Calif., Allan Hancock Found., Los Angeles, 828 pp.

Hedgpeth, Joel W., 1957.
Treatise on Marine Ecology and Paleo Ecology. Vol. I, Memoir 67, Geol. Soc. Amer., 1296 pp.

Hedgpeth, Joel W., 1962.
Seashore Life of the San Francisco Bay region and the Coast of Northern California. Univ. Calif. Press, Berkeley, Calif. Nat. Hist. Guide 9, 136 pp.

Hedgpeth, Joel W., and Sam Hinton, 1961.
Common seashore life of southern California. Naturegraph Co., Healdsburg, California 95448.

Hinton, Sam, 1969.
Seashore Life of Southern California. Univ. Calif. Press, Berkeley, Calif. Nat. Hist. Guide 26, 181 pp.

Johnson, Myrtle E., and Harry J. Snook, 1927.
Seashore Animals of the Pacific Coast. Dover Pub., N.Y., 659 pp.

Keen, A. Myra, 1963.
Marine Molluscan Genera of Western North America. Stanford Univ. Press, Palo Alto, 126 pp.

Keen, A. Myra, and J. C. Pearson, 1952.
Illustrated key to the West North American gastropod genera, Stanford Univ. Press, 34 pp.

Keep, Josiah, 1935.
West Coast Shells. (rev. by Joshua L. Bailey Jr.), Stanford Univ. Press, Palo Alto, 350 pp.

Limbaugh, Conrad, 1955.
Fish life in the kelp beds and the effects of kelp harvesting. Univ. Calif. Inst. Mar. Res., IMR Ref. 55-9, 158 pp.

MacGinitie, George E. and Nettie MacGinitie, 1949.
Natural history of marine animals. McGraw-Hill, London, 473 pp.

McLean, James H., 1962.
Sublittoral ecology of kelp beds of the ocean coast area near Carmel, Calif. Biol. Bull., Vol. 122, No. 1, pp. 95-114.

McLean, James H., 1969.
Marine shells of southern California. Los Angeles County Museum of Natural History, Exposition Park, Los Angeles, California 90007, Science Series 24, Zoology No. 11, 104 pp.

Mertens, Lawrence E., 1970.
In-water photography. Wiley-Interscience, N.Y., 391 pp.

Miller, Daniel J., and Robert N. Lea, 1972.
Guide to the coastal marine fishes of California. Dept. Fish and Game, Sacramento, Fish Bull. 157, 235 pp.

North, Wheeler J., 1964.
Ecology of the rocky nearshore environment in southern California and possible influences of discharged wastes. 1st Intl. Conf. Wtr. Pol. Res., Vol. 3, Pergamon Press, pp. 247-262.

Oldroyd, Ida S., 1957.
The marine shells of the west coast of North America, Stanford Univ. Publ. Geol. Sci. (2) Parts 1-3, pp. 1-941.

Orr, Robert T., 1972.
Marine mammals of California. Univ. Calif. Press, 64 pp.

Osburn, Raymond C., 1950.
Bryozoa of the Pacific coast of America. Part 1. Cheilostomata-Anasca. Allan Hancock Pac. Exped., Vol. 14, pp. 1-269.

Osburn, Raymond C., 1952.
Bryozoa of the Pacific coast of America. Part 2. Cheilostomata-Ascophora. Allan Hancock Pac. Exped., Vol. 14, pp. 271-611.

Osburn, Raymond C., 1953.
Bryozoa of the Pacific coast of America. Part 3. Cyclostomata, Ctenostomata, Entoprocta, and Addenda. Allan Hancock Pac. Exped., Vol. 14, pp. 613-841.

Pequegnat, Willis E., 1964.
The epifauna of a California siltstone reef. Ecology, Vol. 45, pp. 272-283.

Reish, Donald J., 1968.
Marine life of Alamitos Bay. Seaside Printing Co., Long Beach, Calif. (distrib. by 49er Shops Inc., Calif. State Univ. at Long Beach, Long Beach, California 90804), 92 pp.

Ricketts, Edward E. and Jack Calvin, 1968.
Between Pacific tides. 4th Ed. revised by J. W. Hedgpeth. Stanford Univ. Press, 614 pp.

Roedel, Phil M., 1953.
Common Ocean Fishes of the California coast. Dept. Fish and Game, Fish Bull. 91, Sacramento, 184 pp.

Schmitt, Waldo L., 1921.
The Marine Decopod Crustacea of California. Univ. Calif., Pub. Zool., Vol. 23, pp. 1-470.

Shepard, Francis P., 1963.
Submarine geology. Harper & Brothers, New York, 2nd ed., 348 pp.

Shepard, Francis P., and Robert F. Dill, 1966.
Submarine Canyons and Other Sea Valleys. Rand McNally, Chicago, 381 pp.

Smith, Gilbert M., 1944.
Marine algae of the Monterey Peninsula, Calif. Stanford Univ. Press, Stanford, Calif., 622 pp. (Supplement by G. J. Hollenberg and I. A. Abbott added in 1966).

Smith, Ralph I., and James T. Carlton, editors, 3rd rev. ed., 1974.
Light's Manual: Intertidal invertebrates of the central California coast. Univ. Calif. Press, Berkeley, 446 pp.

Turner, Charles H., Earl E. Ebert, and Robert R. Given, 1964. An ecological survey of a marine environment prior to installation of a submarine outfall. California Fish and Game, Vol. 50, pp. 176-188.

Turner, Charles H., Earl E. Ebert, and Robert R. Given, 1965. Survey of the marine environment offshore of San Elijo Lagoon, San Diego County. Calif. Fish and Game, Vol. 51, No. 2, pp. 81-112.

Turner, Charles H., Earl E. Ebert, and Robert R. Given, 1966. The marine environment offshore of Point Loma, San Diego County. San Diego Reg. Wtr. Qual. Cont. Bd., San Diego, 41 pp.

Turner, Charles H., Earl E. Ebert, and Robert R. Given, 1966. The Marine Environment in the Vicinity of the Orange County Sanitation Districts' Ocean outfall. Calif. Fish and Game Quarterly, Vol. 52, pp. 28-48.

Van Name, Willard G., 1945.
The North and South American Ascidians. Bull. Amer. Mus. Nat. Hist., Vol. 84, 476 pp.

LIST OF USEFUL JOURNALS

Aquarius: National Association of Skin Diving Schools, P.O. Box 7666, Long Beach, California 90807.

Biological Bulletin: Marine Biological Laboratory, Woods Hole, Massachusetts 02543.

California Fish and Game Quarterly: State Fisheries Laboratory, Dept. of Fish and Game, 350 Golden Shore, Long Beach, California 90802.

Journal of Experimental Marine Biology and Ecology: North-Holland Pub. Co., P.O. Box 211, Amsterdam, Holland.

Limnology and Oceanography: Dept. of Zoology, Univ. of Wisconsin at Milwaukee, Milwaukee, Wisconsin 53211.

Marine Biology: Springer-Verlag, 175 5th Ave., New York, N.Y. 10010.

Marine Technology Society Journal: Marine Technology Society, 1730 M St., N.W., Suite 412, Washington, D.C. 20036.

Oceans: Oceanic Society, 1255 Portland Place, Boulder, Colorado 80302.

Pacific Science: University Press of Hawaii, 535 Ward Ave., Honolulu, Hawaii 96814.

Sea Frontiers: International Oceanographic Foundation, 1 Rickenbacker Causeway, Virginia Key, Miami, Florida 33149.

Skin Diver Magazine: P.O. Box 3295, Los Angeles, California 90028.

Underwater Naturalist: American Littoral Society, Sandy Hook, Highlands, New Jersey 07732.

Veliger: California Malacozoological Society Inc., 905 Strangler Fig Lane, Sanibel, Florida 33957.

INDEX